T0301789

Unexpected Expectations

Unexpected Expectations

The Curiosities of a Mathematical Crystal Ball

Leonard M. Wapner

CRC Press
Taylor & Francis Group
Boca Raton London New York

CRC Press is an imprint of the
Taylor & Francis Group, an **informa** business

AN A K PETERS BOOK

CRC Press
Taylor & Francis Group
6000 Broken Sound Parkway NW, Suite 300
Boca Raton, FL 33487-2742

© 2012 by Taylor & Francis Group, LLC
CRC Press is an imprint of Taylor & Francis Group, an Informa business

No claim to original U.S. Government works

Version Date: 20120202

International Standard Book Number: 978-1-56881-721-7 (Hardback)

Library of Congress Cataloging-in-Publication Data

Wapner, Leonard M., 1948-
  Unexpected expectations : the curiosities of a mathematical crystal ball / Leonard M. Wapner.
    p. cm.
  "An A K Peters book."
  Summary: "Mathematical expectation or expected value represents the long-term average numerical outcome to an experiment performed a large number of times. Routinely used in the physical sciences, business, and economics, mathematical expectation has also been used to calculate strategies in games of chance and even to justify the belief in God. How can this expression, which is trivial to calculate, have such broad applications and at the same time yield unexpected irresolvable paradoxes? In an easily accessible presentation, this book explores these puzzling and entertaining mysteries"-- Provided by publisher.
  Includes bibliographical references and index.
  ISBN 978-1-56881-721-7 (hardback ; acid-free paper)
  1. Mathematical recreations. I. Title.

QA95.W34 2012
519.2--dc23   2011052564

Visit the Taylor & Francis Web site at
http://www.taylorandfrancis.com

and the CRC Press Web site at
http://www.crcpress.com

For my mom,
Lea Wapner

• • • • • •

For my wife,
Mona Wapner

Table of Contents

Acknowledgments   xi

The Crystal Ball   xiii

1. Looking Back   1
  Beating the Odds: Girolamo Cardano 3
  Vive la France: Blaise Pascal and Pierre de Fermat 6
  Going to Press: Christiaan Huygens 8
  Law, but No Order: Jacob Bernoulli 10
  Three Axioms: Andrei Kolmogorov 13

2. The ABCs of E   19
  The Definition of Probability 20
  The Laws of Probability 22
  Binomial Probabilities 31
  The Definition of Expected Value 32
  Utility 35
  Infinite Series: Some Sum! 37
  Appendix 39

3. Doing the Right Thing   41
  What Happens in Vegas 41
  Is Insurance a Good Bet? 45
  Airline Overbooking 47

Composite Sampling … 51
Pascal's Wager … 54
Game Theory … 56
The St. Petersburg Paradox … 63
Stein's Paradox … 64
Appendix … 68

4. Aversion Perversion   71
Loss Aversion … 72
Ambiguity Aversion … 75
Inequity Aversion … 78
The Dictator Game … 78
The Ultimatum Game … 80
The Trust Game … 80
Off-Target Subjective Probabilities … 82

5. And the Envelope Please!   91
The Classic Envelope Problem: Double or Half … 92
The St. Petersburg Envelope Problem … 94
The "Powers of Three" Envelope Problem … 95
Blackwell's Bet … 97
The Monty Hall Problem … 99
Win-Win … 104
Appendix … 105

6. Parrondo's Paradox: You *Can* Win for Losing   109
Ratchets 101 … 109
The Man Engines of the Cornwall Mines … 110
Parrondo's Paradox … 112
Reliabilism … 114
From Soup to Nuts … 115
Parrondo Profits … 116
Truels: Survival of the Weakest … 117
Going North? Head South! … 120
Appendix … 123

7. Imperfect Recall   127
The Absentminded Driver … 128
Unexpected Lottery Payoffs … 131
Sleeping Beauty … 134
Applications … 137

8. Non-zero-sum Games: The Inadequacy of Individual Rationality   141
Pizza or Pâté … 142
The Threat … 145

Chicken: The *Mamihlapinatapei* Experience 146
The Prisoner's Dilemma 154
The Nash Arbitration Scheme 163
Appendix 167

9. Newcomb's Paradox 169
Dominance vs. Expectation 170
Newcomb + Newcomb = Prisoner's Dilemma 174

10. Benford's Law 177
Simon Newcomb's Discovery 178
Benford's Law 179
What Good Is a Newborn Baby? 184
Appendix 189

Let the Mystery Be! 191

Bibliography 193

Index 199

Acknowledgments

T his book is inspired by family, friends, students, colleagues, and authors too numerous to list. I begin by thanking Martin Gardner for all the years of recreational mathematics he has given us. His passing is a great loss to all who study, teach, and enjoy mathematics.

I thank Phil Everson of Swarthmore College, Robert Vanderbei of Princeton University, and my good friend Paul Wozniak of El Camino College for clarifying various mathematical issues.

Several illustrations appearing in this book were created using Peanut Software, a wonderful collection of nine programs written by Rick Parris of Phillips Exeter Academy, Exeter, New Hampshire. The nine programs are freely distributed, and I thank Rick for his generosity.

Two years ago my friend Michael Carter phoned with the question, "What do you know about Benford's law?" I thank him for the question and the answer now appears as Chapter 10 of this book.

I wish to especially thank Jim Stein, Professor of Mathematics at California State University, Long Beach, California. Jim is the author of four popular mathematics books, and we've had many good discussions of topics in this book while choking down BLTs at a local coffee shop. I'm grateful for his assistance, from the original proposal to the final manuscript. One of our discussions involved the James-Stein theorem, a discussion of which appears in Chapter 3. I had some questions and, after all, who better to ask about the James-Stein theorem than James Stein himself? As it turns out, and as Jim will readily admit, he is neither the James nor the Stein of the James-Stein theorem. Coincidentally, Jim and Willard James (the real James of James-Stein) were colleagues at CSULB many years ago. And, to my surprise, I've come to find out that Willard James was one of my professors. As a student, I never made the connection between Willard James and the James-Stein theorem. It was while choking down one of those BLTs that I put it all together. Small world!

Finally, I wish to thank Klaus Peters, Charlotte Henderson, and all those at A K Peters, Ltd. for turning my manuscript into a book. There may be advantages to electronic, digital publications over hardcover and paperback books; but I'm saddened to think that some day books may no longer exist as we now know them. Will our grandchildren turn pages? What will become of bookmarks and bookcases? Will they only exist *in the cloud*, as amorphous electronic entities? We may just have to get used to the idea. If nothing else, expect paper cuts to decline.

The Crystal Ball

The more unpredictable the world becomes,
the more we rely on predictions.
                                        —Steve Rivkin

I n 1657 the Dutch mathematician and scientist Christiaan Huygens published *De Ratiociniis in Ludo Aleae* (Calculating in Games of Chance), a work generally regarded as the first printed textbook of probability theory (see Figure 1). In so doing he gave the sciences a statistical crystal ball—a means of predicting, with some degree of reliability, that which is uncertain.

Mathematical expectation or *expected value* is the long-term average numerical outcome to an experiment performed a large number of times. It's denoted as

$$E = \Sigma xp$$

and can be applied, in some sense, to one-time (single-shot) experiments. This unassuming expression is both simple in appearance and easy to calculate, often requiring nothing more than multiplication and addition. (The above expression will be explained in detail in Chapter 2.) Mathematical expectation has served the sciences well for more than three centuries with applications in such diverse fields as game theory, quantum mechanics and the forensic sciences. It was even used by Blaise Pascal to rationalize one's belief in God! Though useful, its very definition is paradoxical in that its calculation requires the laws of chance. The *laws of chance*? Isn't chance, by definition, not governed by laws? Can there truly be a *science of the uncertain*?

Figure 1. Christiaan Huygens (1629–1695). (From Leiden University Library, Print Room, PK-P-108.246.)

By popular standards there may be other, far more significant mathematical expressions. In *Five Equations That Changed the World*, Michael Guillen offers these five examples [Guillen 95]:

1. Isaac Newton's Universal Law of Gravity: $F = \frac{GMm}{d^2}$;
2. Daniel Bernoulli's Law of Hydrodynamic Pressure: $P + \frac{1}{2}\rho v^2 = $ constant;
3. Michael Faraday's Law of Electromagnetic Induction: $\nabla \times E = -\frac{\partial B}{\partial t}$;
4. Rudolf Clausius' Second Law of Thermodynamics: $\Delta S_{\text{Universe}} > 0$ (inequality);
5. Albert Einstein's Theory of Special Relativity Equation: $E = mc^2$.

All are worthy of Guillen's acknowledgement. Mathematical expectation may not be as glamorous, but its applicability should not be underestimated. Elementary textbooks of probability and statistics include many examples.

This book showcases *unexpected* mathematical expectations—those that surprise us in one way or another. We begin by noting that the very existence of such an expression is surprising enough. For someone unfamiliar with the subject matter, the power and diversity of its applications are certainly unexpected. But things get more bizarre when we discover mathematical paradoxes and contradictions associated with straightforward applications of the concept. These contradictions are highly unexpected and may lead one to reconsider the legitimacy of the trusted applications. These curiosities are the core of this book. Readers will be intrigued; some may be frustrated. This is the nature of paradox.

The book begins with a brief history of probability theory, a relatively recent development considering thousands of years of mathematical progress. Advances in science and technology are often initiated by less-than-glamorous aspects of human nature. The uniquely human quality of making war has produced advances in aviation, communication, and GPS navigation. Some narcotics and other recreational drugs have proven to be of medicinal value. The pornography industry may be responsible for the VHS video format succeeding over its rival Beta, and may prove to be a deciding factor in the current Blu-ray versus HD-DVD rivalry. Probability theory was born from the desire to develop gambling strategies, a vice of sorts. Recreational greed!

Chapter 2 covers the rudiments—the *ABCs of E*. It includes the basic laws of probability and the definition of mathematical expectation (expected value). Chapter 3 surveys routine applications of *E*, with a surprise or two even for readers familiar with the concept.

Remaining chapters cover unexpected results associated with mathematical expectation. Chapter 4 is where psychology meets mathematics—$\Psi + E$. Here we see we're not always as rational as we believe ourselves to be. In Chapter 5 we examine a class of expected value paradoxes which I collectively refer to as *envelope problems*. One must choose one of two envelopes so as to maximize the expected value of the contents. More generally, the decision is between one of two possible courses of action. It's not as straightforward as one might imagine, and the reader should expect the unexpected.

Are stock market losses getting you down? Do you consider yourself an expert at picking losing stocks or making other bad bets? If so, then take heart! Chapter 6 offers hope to losers. Chapter 7 presents the problems associated with *imperfect recall*. The original problem from the seminal paper [Piccione and Rubinstein 97] is given, followed by popular variations.

Chapter 8 gives a brief discussion of non-zero-sum games, where again we encounter the inadequacy of individual rationality. Attention is given to the classic game of chicken and the prisoner's dilemma, both of which have real-world manifestations.

Newcomb's paradox, presented in Chapter 9, may be considered the greatest philosophical paradox of free will. It is similar to the envelope problems presented in Chapter 5. But to choose optimally, one must consider the age old *free will versus determinism* debate. It is more a philosophical than mathematical problem and deserves its own chapter.

The letter *E* (or *e*) denotes many mathematical and physical quantities. Lower case *e* is commonly used for eccentricity (the elongation of conic section curves such as ellipses and hyperbolas) and also denotes the base of the natural logarithm function ($e \approx 2.7$). Upper case *E* may represent energy (as in $E = mc^2$), voltage (Ohm's law: $E = IR$), and mathematical expectation. In Chapter 10 we see how mathematical expectation can be applied to the forensic sciences and serve as evidence in our courts. *E* for evidence! A recently discovered phenomenon,

Benford's law, can be used forensically as evidence of fraud. The law was recently used to detect anomalies in the 2009 Iranian presidential election vote count. The mathematical discussions to follow typically involve little more than basic (high school) algebraic operations. On those few occasions where a higher level of mathematics is required, such as calculus or Markov chains, the discussion appears in the chapter's appendix so as not to deter those readers unfamiliar with such subjects.

I sincerely hope the reader will appreciate the applications of this statistical crystal ball, along with its magic and mystery. Mathematics is an art, perhaps more so than a science. With or without application, its magic and mystery are endless. Legendary mathematician John von Neumann once noted, "In mathematics you don't understand things. You just get used to them." As author it is my hope to bring readers well beyond tolerance; the mysteries of mathematics are to be embraced. Shakespeare writes, "Expectation is the root of all heartache." He must not have read this book!

—Leonard M. Wapner
Division of Mathematical Sciences
El Camino College

Looking Back

*If you would understand anything,
observe its beginning and its development.*
—Aristotle

During every waking moment we make decisions with uncertain outcomes. We intuitively weigh the value (payoff) of these outcomes, consider the risks, and make our decisions. In some instances the payoff is immediate, such as that of a wager at a casino table. Other decisions, such as deciding how to invest a significant sum of money, may require decades before the outcome is known and can be evaluated. And then there are those apparently insignificant decisions we make daily that actually have a greater impact on our lives than we can imagine when the decision is made.

On June 18, 1977, I boarded the 10:02 am train from Copenhagen to Hamburg. I was tired, traveling with a pack on my back, and walked through the train searching for an empty compartment. I bypassed a compartment with only one passenger, hoping to find a compartment that I would have all to myself. Other compartments were more crowded, and I made the decision to return to the compartment occupied by the single passenger. At the time, I gave the decision little thought. As it turns out, this may have been the most significant decision of my life. The passenger was Mona and we are still *traveling together*, now happily married for over 35 years. Our daughter, Kirsty, and many other happy events in our lives would not have occurred had I chosen any other compartment on the train.

A single decision today may have immediate payoffs or far reaching, unknown consequences down the road. The phenomenon of long-term dramatic consequences associated with an apparent insignificant decision is similar to the *butterfly effect*, where small physical perturbations yield unpredictable major out-comes. According to famed meteorologist Edward Lorenz [Lorenz 93, p. 14], the origin of this expression may be Lorenz's paper entitled "Does the Flap of a But-terfly's Wings in Brazil Set Off a Tornado in Texas?"

If outcomes can be quantified, or at least ranked, then a rational individual makes decisions so as to maximize the value of the outcome. But, if the outcomes are chance or highly unpredictable events, then the optimal decision becomes problematic. Prior to the sixteenth century, decisions having chance outcomes were based on intuition or considered fate, as chance events were believed to be governed by the gods or other supernatural powers, if they were governed at all. Consideration of chance events and the likelihood of their occurrence was most often related to games and gambling. At this point, a distinction must be made between an awareness of chance and an understanding of chance at a level on which a mathematical theory of probability is developed. Awareness of the con-cept dates back six millennia, but it was not until the European Renaissance that the theory of probability was born.

The standard cubical die has served as a common randomizer for today's board games and other games of chance. This verse of an American folksong de-scribes dice as *bones*, which are, in fact, the original form of dice:

> Roll, roll, roll dem bones,
> Roll dem in de square,
> Roll dem on de sidewalk,
> Streets or anywhere!
> We roll dem in de morning,
> We roll dem in de night.
> Oh, we roll dem bones de whole day long
> When de cops are out of sight!

The *astragalus* (or *talus*) bone, taken from the heel of goats, sheep, or other hooved animals, is cubical and relatively symmetrical. Evidence from Egyptian tomb paintings and other archeological evidence indicate that astragali were used in ancient civilizations as game pieces, counters, a medium of exchange, and ran-domizers in games. The astragali were thrown as dice are rolled today. The out-come was given by the orientation of the astragali. Historical records date this back to ancient civilizations—the Sumerians, Babylonians, Assyrians, Greeks, and Romans—indicating an awareness of chance, but nothing in the way of un-derstanding or theory. The fact that the astragali were not perfectly symmetrical and not all alike undoubtedly hindered empirical analysis.

The odds-on favorite of being the first civilization to develop the theory of being an odds-on favorite would have been Ancient Greece. Greek vases show paintings of men tossing astragali. In Greek mythology, there is the story of the three brothers, Zeus, Poseidon, and Hades, throwing dice to win a share of the universe. Zeus wins the heavens, Poseidon the seas, and Hades the underworld (hell). The Greeks certainly had a notion of chance, being fond of games and gambling. And, they were making significant contributions to western culture with respect to mathematics, logic, philosophy, science, and the arts. But, in actuality, they contributed little to the development of a theory of probability. Socrates (469–399 BC) defined εικος (*eikos*) as a word meaning *probable* or *likely*. But with a number system more suited to record than calculate, the Greeks were not able to advance a mathematical theory of probability.

Ancient Rome is known for its military conquests and advances in architecture and engineering—not so much for advances in science and mathematics. (Can you name a single Roman mathematician?) Like the number system of Ancient Greece, the Roman number system was, and remains today, not well suited for calculation. It would not be until 500 AD that the Hindus would develop a number system allowing for calculations similar to those of today. Nevertheless, the Romans should be considered the first to quantify, albeit in a minimal sense, probability. Cicero (106–43 BC) used the word *probabilis*, being aware that among chance events, some were more likely to occur than others. He writes, "probability is the very guide of life" [Cicero 63, p. 53].

James Franklin, in his book *The Science of Conjecture*, gives an interesting passage of text extracted from a collection of old papal letters and quotations, dated about 850 AD [Franklin 01, p. 13–14]:

> A bishop should not be condemned except with seventy-two witnesses ... a cardinal priest should not be condemned except with forty-four witnesses, a cardinal deacon of the city of Rome with thirty-six witnesses, a subdeacon, acolyte, exorcist, lector, or doorkeeper except with seven witnesses.

Franklyn writes [Franklin 01, p. 14], "It is the world's first quantitative theory of probability. Which shows why being quantitative about probability is not necessarily a good thing."

Roman medieval law mentions the concept *semiplena probatio* (half-proof), a level of evidence somewhere between suspicion and full proof needed to convict. In some cases, the degree of likelihood of truth was further subdivided. Similar concepts existed in seventeenth-century English law and nineteenth-century Scottish law. Today, in US civil law, the plaintiff wins if the jury believes the probability is greater than 50 percent that the defendant was negligent. In US criminal courts, the burden of proof is "beyond reasonable doubt," quantified as a 98–99 percent certainty of guilt.

Two great western civilizations, the Greeks and the Romans, failed to develop anything in the way of mathematical probability. Fast forward 1,500 years to the European Renaissance, where the theory of probability was born.

Beating the Odds: Girolamo Cardano

The first written formulation of a theory of chance was that of Girolamo Cardano (1501–1576). Leading a colorful life as a physician, scientist, and mathematician, he may best be known as an eccentric gambler and author of *Liber de Ludo Aleae* (The Book on Games of Chance), written in 1525 but not published until 1663. From conception to death, Cardano was a gambler, beating the odds in every sense. He was born frail and out of wedlock in Pavia (now Italy), despite repeated attempts by his mother, Chiara, to abort the fetus. Months later he survived the bubonic plague, though it killed his nurse, three half brothers, and well over 25 percent of the European population. In later years his eldest son was executed for murder and his youngest was exiled as a thief. At the age of 69 Cardano was arrested and briefly jailed as a heretic for having cast the horoscope of Jesus Christ. Cardano died in Rome on September 20, 1576, with some suggesting that he had predicted he would die on this day and starved himself for the preceding three weeks so that his prediction would come true.

Growing up Cardano studied science and mathematics. In 1516, he decided to study medicine, ultimately becoming one of Europe's most prominent physicians. It was as a medical student that Cardano became interested in gambling (dice, chess, and card games), using winnings as his primary means of income. In his autobiography, he frankly acknowledges the addiction [Cardano 02, p. 66]:

> Peradventure in no respect can I be deemed worthy of praise; for so surely as I was inordinately addicted to the chess-board and the dicing table, I know that I must be deserving of the severest censure. I gambled at both for many years, at chess more than forty years, at dice about twenty-five; and not only every year, but—I say it with shame—every day, and with the loss at once of thought, of substance, and of time.

The combination of his passion for gambling and exceptional mathematical talent gave way to his book on games of chance and history's first written treatise on the laws of chance. The book is essentially a gambling manual, with much written about dice games, to which he was introduced by his sons. Of interest is Chapter 14 entitled "On Combined Points." Here Cardano introduces what is now known as a *sample space*, the set of all possible simple outcomes associated with some random process. (Details are given in Chapter 2.) If the outcomes are equally likely and if some are considered favorable, then Cardano notes that the likelihood of a favorable outcome is equal to the proportion of favorable outcomes.

With reference to rolling a pair of six-sided dice, Cardano writes of the likelihood of rolling a 1, 2, or 3 on either of both dice [Cardano 53, pp. 200–202]:

> If, therefore, someone should say, "I want an ace, a deuce, or a trey," you know that there are 27 favorable throws, and since the circuit is 36, the rest of the throws in which these points will not turn up will be 9; the odds will therefore be 3 to 1.

Cardano's expression "circuit is 36" refers to the sample space, which for a pair of dice consists of 36 equally likely outcomes. (For each of the six possible outcomes

(1,1), (1,2), (1,3), (1,4), (1,5), (1,6),
(2,1), (2,2), (2,3), (2,4), (2,5), (2,6),
(3,1), (3,2), (3,3), (3,4), (3,5), (3,6),
(4,1), (4,2), (4,3), (4,4), (4,5), (4,6),
(5,1), (5,2), (5,3), (5,4), (5,5), (5,6),
(6,1), (6,2), (6,3), (6,4), (6,5), (6,6)

Table 1.1. The 27 ways of rolling an ace, deuce, or trey out of all possible results of rolling a pair of dice.

for one die, there exist six possible outcomes for the other.) The "27 favorable throws," corresponding to the 27 ways a 1, 2, or 3 can appear on either of both dice, are shown in bold in Table 1.1.

He summarizes,

> So there is one general rule, namely, that we should consider the whole circuit, and the number of those casts which represents in how many ways the favorable result can occur, and compare that number to the remainder of the circuit, and according to that proportion should the mutual wagers be laid so that one may contend on equal terms.

Cardano's correct analysis introduces today's notion of probability as a fraction p, $0 \le p \le 1$, of the desired results over the total possible results. For Cardano's example above, a favorable result occurs 27 out of 36 times and the probability of a favorable result is $27/36 = 3/4$.

Cardano's short book was groundbreaking, and successful. Beyond formulating probability as a proportion (fraction), he proposes, without proof, the law of large numbers (discussed later in this chapter) and derives some other fundamental laws of modern probability theory (Chapter 2). There are some shortcomings. By modern standards, there was relatively no mathematical symbolism and principles are given in terms of examples. Some errors are made, which he notes; yet, there are no corrections. At times his writing is vague and hard to follow. Dispersed throughout the book are nontechnical, sometimes amusing comments about the nature of gambling and character of the players [Cardano 53, pp. 186–188]: for example,

> In my own case, when it seemed to me after a long illness that death was close at hand, I found no little solace in playing constantly at dice. However, there must be moderation in the amount of money involved; otherwise, it is certain that no one should ever play.
> To these facts must be added that gambling arouses anger and disturbs the mind, and that sometimes a quarrel flares up over money, a thing which is disgraceful, dangerous, and prohibited by law.

Your opponent should be of suitable station in life; you should play rarely and for short periods, in a suitable place, for small stakes, and on suitable occasions, as at a holiday banquet.

Play is a very good test of a man's patience or impatience. The greatest advantage in gambling comes from not playing at all. But there is a very great utility in it as a test of patience, for a good man will refrain from anger even at the moment of rising from the game in defeat.

Cardano was not the first, nor the last, to consider the unsolved *problem of the points*, also known as the *problem of division of stakes*. How should prize money be divided between two players if it is necessary to terminate a contest before its completion, when the players have only partial scores? This is equivalent to asking the probability of each player winning at every possible stage of the game. The problem had its origins in the Dark Ages and appears in *Summa de arithmetica, geometria, proportioni e proportionalità*, published in 1494 by the Franciscan monk Luca Pacioli (1445–1517). Born in Tuscany, Pacioli was well educated in mathematics, with both teaching experience and a background in business. He wrote books about arithmetic and algebra which, for the most part, contained nothing original. Today he is called the "father of accounting" because of his detailed descriptions of the double-entry system in use today.

Cardano, like Pacioli, was unable to correctly solve the problem of the points, despite the fact its solution is trivial in the context of today's theory of probability. As a simple version, consider two tennis players of equal ability playing a match in which the winner is the first to win three sets. Assume the match is interrupted and must be terminated when one player leads the other by the score 2-1. How should the prize money be divided? For Pacioli, Cardano and others, this was no simple problem. Would it be proper to divide the stakes evenly? This seems unfair as the player with the score of 2 would be more likely to win if the match had continued. Why not award the leading player two thirds of the stakes as he has scored two thirds of the points when the match was terminated? This too is illogical since, by this convention, a match terminated at 1-0 would award 100 percent of the stakes to the leading player, even though it is far from certain this player would win the match had it continued.

As it turned out, the problem of the points was an embryo awaiting development. Though in essence a moral question of fairness, it requires a simple, but essential, quantification of chance in order to be solved. And, it introduces the notion of mathematical expectation, the subject of this book. Once the problem was passed on to seventeenth-century French mathematicians, the problem was solved and the theory of probability continued to develop.

Vive la France: Blaise Pascal and Pierre de Fermat

Throughout the seventeenth and eighteenth centuries, French mathematicians including Fermat, Pascal, de Moivre, Laplace, and Poisson took the lead in advancing the rapidly developing theory of probability. Ironically, it was a chance event that got the ball rolling. The French writer and socialite Antoine Gambaud (1607–1684), known by the title Chevalier de Méré, was a professional gambler, but not an expert mathematician. By chance, he was introduced to famed French scientist and mathematician Blaise Pascal (1623–1662), and de Méré took the opportunity to ask for Pascal's help in solving certain gambling related problems, one of which was the problem of the points. In 1654 Pascal passed de Méré's questions on to another French mathematician, Pierre de Fermat (1601–1665), considered by many as the world's greatest amateur mathematician. A brief but significant correspondence between the two mathematicians followed that forged the basic mathematical concepts of probability. Both Pascal and Fermat solved the problem of the points, but by different methods. Pascal's approach uses Cardano's idea of the totality of equally likely outcomes; a simplified version appears here.

Let the tennis players be named Hedrick and Taylor. Hedrick leads with the score at 2-1. The first player to reach 3 wins the match. So, at most, two additional sets must be played to conclude the match. Assuming the players are of equal ability, the scoring is equivalent to tossing a coin: whenever it turns up heads (H), Hedrick wins a set; when it comes up tails (T), Taylor wins a set. There are four possible outcomes for the two possible remaining sets—HH, HT, TH, and TT. (In practice, the match ends once either player wins the third set. So, of the four possibilities listed, HH and HT would not actually occur as play would be terminated when Hedrick wins the third set. But, when considering all possible outcomes for playing the two possible remaining sets, we must include these two possibilities.) Three of these four equally likely outcomes correspond to H winning the match. With this in mind, the stakes should be divided so that H gets three parts and T gets one part. This is equivalent to stating that Hedrick's probability of winning the match, given that he is leading by the score 2-1, is 3/4.

This solution is significant on several levels. It quantifies probability and firmly establishes the concept of a sample space, introduced by Cardano. The solution implicitly uses the idea of mathematical expectation, given here as a fair division of stakes, obtained from the notion of what the players would expect to win, on average, if play were to continue. And, there were practical applications beyond that of interrupted games. The European Renaissance was a time of commercial growth, as well as a rebirth of learning and culture. Then, as today, lenders would lend money to merchants and both would profit if the merchants were successful. The merchants could expand their business and return a portion of the profits as interest to the lenders. Both parties assumed a share of the risk. In this context, the problem of the points addressed how lenders and merchants should settle their agreement if the venture did not pan out. Some scholars suggest the problem of the points was motivated more by business concerns than those of gambling.

The advances made by Pascal and Fermat with respect to probability survived, but only in the form of their mutual correspondence; neither published them. The conclusions of the two Frenchmen were discussed by other French mathematicians; but, it was a Dutchman who became so intrigued by the correspondence that he proceeded to write the first printed textbook on the subject.

Going to Press: Christiaan Huygens

Christiaan Huygens was born at The Hague in 1629 into a family of wealth and influence. His father, Constantijn, was a statesman and one of the best known poets of the Netherlands. Constantijn was well connected academically, corresponding regularly with French philosophers/mathematicians Marin Mersenne (1588–1648) and René Descartes (1596–1650). His mother, Suzanna Van Baerle, was from one of the richest families in Amsterdam. So, much was expected of the young Christiaan. As a child he was privately tutored, then he studied law and mathematics at the University of Leyden and the College of Orange at Breda.

While visiting Paris in 1655, Huygens learned of the problem of the points and the Pascal–Fermat correspondence. At the time, Huygens never met either of the two Frenchmen, but he was sufficiently intrigued by the subject of their correspondence to continue his own investigations upon returning to the Netherlands. Dutch mathematician and publisher F. van Schooten requested Huygens write up his results for publication. One year later, after a brief correspondence with Pascal and Fermat, Huygens completed *De Ratiociniis in Ludo Aleae* (Calculating in Games of Chance) and submitted the manuscript. Huygens wrote in Dutch, having difficulty finding Latin words for the subject matter. Van Schooten published the Latin translation in 1657; the Dutch version appeared in 1660 as *Van Rekiningh in Spelen van Geluck*. The book is regarded as the first printed textbook of probability theory; yet, by today's standard, Huygens' treatise is remarkably short, only 13 pages if printed in the text format you are now reading.

The work consists of an introductory paragraph referencing the problem of the points, followed by a single postulate where he sets down the notion of mathematical expectation as the value of a gamble. This is Huygens' definition, as given by W. Browne's 1714 English translation of Huygens' book [Huygens 14]:

> As a Foundation to the following *Proposition*, I shall take Leave to lay down this Self-evident Truth: *That any one Chance or Expectation to win any thing is worth just such a Sum, as wou'd procure in the same Chance and Expectation at a fair Lay.* As, for Example, if any one shou'd put 3 Shillings in one Hand, without telling me which, and 7 in the other, and give me Choice of either of them; I say, it is the same thing as if he shou'd give me 5 Shillings; because with 5 Shillings I can, at a fair Lay, procure the same even Chance or Expectation to win 3 or 7 Shillings.

(Today's equivalent interpretation of mathematical expectation is that of a long-term average numerical outcome. Using the above example of choosing a hand containing either 3 or 7 shillings, each equally likely, the long-term average outcome would be that of getting 5 shillings. A third interpretation of expected value is that of a proportion of stakes, as given in the problem of the points.)

Fourteen propositions then follow. Propositions I–III give Huygens' technique for calculating mathematical expectation. These three propositions, with Huygens' justification of the first, are again presented as written by Huygens and translated by Browne [Huygens 14].

Proposition I

If I expect a or b and have an equal chance of gaining either of them, my Expectation is worth $(a + b)/2$.

For most readers, this proposition will appear self evident, not requiring any justification. But being the first proposition of the first book of its kind, Huygens justifies the proposition, referencing his opening postulate [Huygens 14]:

> To trace this Rule from its first Foundation, as well as demonstrate it, having put x for the value of my Expectation, I must with x be able to procure the same Expectation at a fair Lay. Suppose then that I play with another upon this Condition, That each shall stake x, and he that wins [say by the toss of a fair coin] give the Loser a. 'Tis plain, the Play is fair, and that I have upon this Agreement an even Chance to gain a, if I lose the Game; or $2x - a$, if I win it: for then I have the whole stake $2x$, out of which I am to pay my Adversary a. And if $2x - a$ be supposed to equal b, then I have an even Chance to gain either a or b. Therefore putting $2x - a = b$, we have $x = (a + b)/2$, for the Value of My Expectation.

To be clear, imagine yourself and another player each wagering $(a + b)/2$ on the outcome of a coin toss. It is agreed that the winner of the toss, after collecting the stakes, returns an amount a to the loser. By symmetry, the game is fair. If you lose the coin toss, you walk away with a, returned to you by the other player, as agreed. If you win the toss, you walk away with $2(\frac{a+b}{2}) - a = b$. So, for both you and the other player, there is equal likelihood of coming away with a or b, procured with $(a + b)/2$.

Propositions II and III are given below. Huygens' justifications are similar to that used for Proposition I and are omitted here.

Proposition II

If I expect a, b, or c, and each of them be equally likely to fall to my Share, my Expectation is worth $(a + b + c)/3$.

Proposition III

If the number of Chances I have to gain a be p, and the number of Chances I have to gain b, be q. Supposing the Chances equal; my Expectation will then be worth $\frac{ap+bq}{p+q}$.

Propositions IV–IX deal specifically with the problem of the points. Propositions X–XIV involve dice. The book concludes (as do most chapters in today's mathematics textbooks) with five exercises for the reader. The short but significant treatise of Huygens was well received and for fifty years remained the single best introduction to the developing theory of mathematical probability.

Today Huygens is better known for his work in astronomy and other natural sciences. With an interest in lenses and telescopes, he invented a new way of grinding and polishing lenses, then used these lenses to discover the true nature of the rings of Saturn. Astronomy and time keeping go hand in hand, and his interests led him to develop the first pendulum clock, significantly improving the ability to accurately measure time. Precise timekeeping was also necessary for navigation at sea in order to determine longitude. (Lattitude can be approximated by celestial navigation.) To this end, Huygens invented the cycloidal pendulum, well suited for measuring time on a rolling sea.

Huygens biographer A. E. Bell describes Huygens as "one of the greatest scientific geniuses of all time" and manages to sum up his broad scientific achievements in one, unusually long sentence [Bell 47, p. 5]:

> A man who transformed the telescope from being a toy into a powerful instrument of investigation, and this as a consequence of profound optical researches; who discovered Saturn's ring and the satellite Titan; who drew attention to the Nebula in Orion; who studied the problem of gravity in a quantitative manner, arriving at correct ideas about the effects of centrifugal force and the shape of the earth; who, in the great work *Horologium Oscillatorium*, founded the dynamics of systems and cleared up the whole subject of the compound pendulum and the tautochrone; who solved the outstanding problems concerned with collision of elastic bodies and out of much intractable work developed the general notion of energy and work; who is rightly regarded as the founder of the wave theory in light, and thus of physical optics—such a man deserves memory with the names of Galileo and Newton, and only the accidents of history have prevented this.

In 1694, Huygens' health began to deteriorate. Suffering from insomnia, pain, and despair, he died in 1695 at his birthplace, The Hague.

Law, but No Order: Jacob Bernoulli

Is it *nature* (heredity) or *nurture* (environment) that has the greater effect on a person's interests, talents, and choice of vocation? The *nature vs. nurture* debate

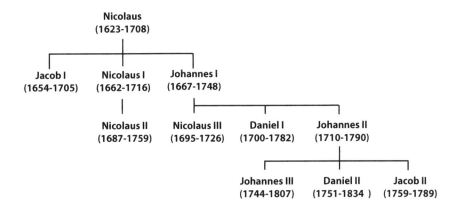

Figure 1.1. The Bernoulli family.

has been around for some time, and evidence exists supporting both sides of the argument. Why would one choose to study mathematics and become a mathematician? Is it something in an individual's genes (nature) that predisposes one for this type of work? As evidence of *apples not falling far from the tree*, there can be no better example than that of the Bernoulli family—with its three generations of famous mathematicians (see Figure 1.1). No family in history has produced more mathematicians.

Fleeing persecution by the Catholics, the Bernoullis escaped Antwerp in 1583 and made it to Basel by way of Frankfurt. Jacob (James, Jacques) Bernoulli (1654–1705), eldest son of Nicolaus senior, was born on Christmas Day, the same year as the Pascal–Fermat correspondence. Jacob, like his brothers Nicolaus and Johannes, was not raised by Nicolaus senior to be a mathematician. Encouraged by his father, Jacob earned a degree in theology at Basel and trained to be a minister of the Reformed Church. He spent the next ten years traveling throughout Europe, developing an interest in astronomy and mathematics, studying the writings of Descartes, and corresponding with German philosopher and mathematician Gottfried Wilhelm Leibniz.

Now a mathematician, Jacob returned to Basel and, in 1687, became chair of mathematics at the University of Basel. He held this position until his death in 1705. During these years he became interested in Huygens' *De Ratiociniis in Ludo Aleae* and began working on problems of chance, such as those posed by Huygens. Now influenced by both Huygens and Leibniz, Jacob combined the ideas of calculus and probability to write *Ars Conjectandi* (The Art of Conjecturing), composing the manuscript over a period of twenty years, the first part of which contains a reprint of Huygens' short work. Jacob died in 1705 before the book's publication. Nicholaus, the son of Jacob's brother Nicholaus, made final edits and in 1713 *Ars Conjectandi* was published.

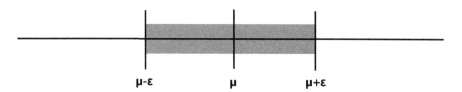

$\mu - \varepsilon$ μ $\mu + \varepsilon$

Figure 1.2. The law of large numbers.

The primary contribution in this book to the theory of probability is Bernoulli's proof of the first limit law of probability, the *law of large numbers*, also known as *Bernoulli's theorem*. Cardano expressed the idea, but it was Bernoulli who produced a formal proof, which is omitted here. The law can be expressed in terms of averages, proportions, or probabilities. In that probabilities are not defined until the following chapter, we delay that form of the law until the following chapter. In simplest terms, the law states that the difference between the average of *n* numbers randomly chosen from a large data set and the true average of the entire data set, tends to decrease as the sample size *n* increases. More formally, let μ represent the average value of some quantity associated with a large population. For any positive value ε, create a symmetric interval centered at μ as depicted in Figure 1.2. Then the likelihood of the sample mean being in this window, between $\mu - \varepsilon$ and $\mu + \varepsilon$ increases as the sample size *n* increases, regardless of the value of ε.

Bernoulli believed that the result was self evident. In *Ars Conjectandi* he writes [Bernoulli 06, p. 328],

> Neither should it escape anyone that to judge in this way concerning some future event it would not suffice to take one or another experiment, but a great abundance of experiments would be required, given that even the most foolish person, by some instinct of nature, alone and with no previous instruction (which is truly astonishing), has discovered that the more observations of this sort are made, the less danger there will be of error. But although this is naturally known to everyone, the demonstration by which it can be inferred from the principles of the art [of conjecturing] is hardly known at all, and, accordingly, it is incumbent upon us to expound it here.

Obvious is a dangerous word to use in mathematics; the theorem's conclusion is obvious, but not its proof. We know there is "safety in numbers," "fifty million Frenchman can't be wrong," and, by Proverbs 11:14 of the King James Bible, "Where no counsel is, the people fall: but in the multitude of counselors there is safety." The law of large numbers is intuitive.

(Bernoulli's theorem is known today as the *weak law of large numbers*. The *strong law of large numbers*, which implies the weak law, asserts that as *n* increases, the average of the sample almost certainly approaches the true mean for the entire data set. The distinction between the two versions will be subtle for some readers. The weak version allows that the difference between the two means can exceed ε infinitely many times as n increases. The strong version asserts that this will almost certainly not occur. The distinction need not concern us.)

The converse of the law of large numbers is of great practical value. So far, we have assumed that the true average for the entire population is known, in which case the average of the sample will approach that of the population as the sample size n increases. But what if we don't know the true average? The converse states that it would be possible to approximate this value, to any desired degree of accuracy and to any specified degree of confidence, by taking a sufficiently large sample and using the calculated result for the sample as an approximation for the true value. The converse is, in fact, true; but, the proof eluded Bernoulli.

As an example of the converse, assume a university's department of student health wishes to estimate the average body weight of its 15,000 male students. The department wants its estimate to be within $\varepsilon = 5$ pounds of the true average of all 15,000 males. For a random sample of $n = 100$ male students, the average weight is 172 pounds. It is likely that they are within 5 pounds of the true average (μ); but, it would be more likely if the sample consisted of more students.

The law of large numbers applies to proportions as well since a proportion is an average of sorts. To estimate the proportion of those male students weighing more than 230 pounds, the department could, as a matter of convenience, select a sample of $n = 100$ male students and find the proportion of this sample exceeding 230 pounds. A larger sample would allow the department to be more confident in its estimation of the true proportion. The sample size is limited, usually due to constraints (finances, time, etc.).

The real significance of the law of large numbers is that it allows us to extract some form of order from chaos. The outcome of a coin toss is random and unpredictable; yet, as the number of tosses increases, the proportion of heads predictably approaches 1/2. It is this predictability that allows us to use mathematical expectation to, in some sense, predict the future outcomes of a random process.

Three Axioms: Andrei Kolmogorov

The development of a mathematical theory is in many ways like that of a child. There is conception, followed by birth, periods of accelerated growth, and milestone events worthy of recognition. The origins of probability can be traced back to the astragali of ancient civilizations. Coaxed by Cardano, Pascal, and Fermat, probability's birth appeared somewhat prematurely with the publication of *De Ratiociniis in Ludo Aleae*. For 200 years following Jacob Bernoulli, mathematicians, notably the French, advanced the theory to what it is today.

Frenchman Abraham De Moivre (1667–1754) authored *The Doctrine of Chances*, a work recognized as one of the first great texts on probability. In his mid-sixties he discovered the bell curve (normal distribution, Gaussian distribution), further developed years later by the German mathematician Carl Friedrich Gauss.

Swiss mathematician Leonhard Euler (1707–1783) contributed to many branches of mathematics, including probability. Highly prolific, he published over 400 papers, more than 350 of which during the last ten years of his life, when he was blind. With respect to probability, Euler's work related to games of chance and, in particular, the analysis of state-run lotteries. Two such papers were written by Euler while working in Berlin for Frederick the Great, King of Prussia, to help implement lotteries designed to pay off war debt.

Pierre-Simon De Laplace (1749–1827), in his book *Théorie Analytique des Probabilités*, extended probability beyond games of chance, developing applications in actuarial mathematics and statistical mechanics.

There have been many other significant contributors—D'Alembert, Poisson, Markov, to name but a few. There is no proper way to end the list, as the theory of probability continues to develop, and will do so for years to come. But, there is one significant event in the theory's development to serve as this chapter's proper conclusion. Russian mathematician Andrei Kolmogorov (1903–1987), in answer to a mathematical challenge by German mathematician David Hilbert, succeeded in axiomatizing the theory, providing it a formal status like that of Euclidean geometry, the Peano axiomatization of the natural numbers, and set theory. This formalization lead to probability theory becoming part of a more general branch of mathematics known as *measure theory*.

Andrei Nikolaevich Kolmogorov was born in Tambov, Russia, in 1903 to unmarried parents. His mother died giving birth to Andrei and his father, exiled for a time, died when Andrei was 16. (The name "Kolmogorov" was not that of Andrei's father; it came from his grandfather Yakov Stepanovich Kolmogorov.) Young Andrei was raised by his aunt, Vera Yakovlena, who adopted him in 1910. In 1920, Kolmogorov entered Moscow State University studying metallurgy, Russian history, and mathematics. Even as an undergraduate, Kolmogorov made a reputation for himself doing mathematical research and producing results in set theory and Fourier series. Then gravitating to mathematics, he continued at Moscow State University, earning his doctorate in 1929.

During his life, Kolmogorov made major contributions to mathematical and scientific branches including topology, set theory, algorithmic complexity theory (Kolmogorov complexity theory), dynamical systems, and classical mechanics. But, here we focus on his axiomatization of probability theory, which gave it the sense of rigor it had been lacking.

What is an *axiomatic system*? An axiomatic system consists of a set of axioms (postulates, first principles, assumed truths) from which additional statements, or theorems, are derived by logical proof. The system may contain both defined and undefined terms. In its entirety, a *mathematical theory* is the set of all axioms of the system, along with all theorems derived from the axioms. We are typically introduced to the concept when taking a first course in Euclidean geometry. Other familiar axiomatic systems include the Peano axiomatization of the natural numbers and axiomatic set theory. An axiomatic system is *consistent* if it does not admit contradictions. An axiomatic system is *complete* if every statement containing the terms of the system can either be proved or disproved within that system. Kurt Gödel's famous 1931 paper shows that any axiomatizable theory involving the natural number system is incomplete; there will always be statements that can neither be proved nor disproved using only natural numbers. (Two non-mathematical examples of non-provable statements would be, "This sentence is false" and "All general statements are false." Both are self-referential statements.) Here is an example of a simple, axiomatic system:

1. There are exactly three members in the band.
2. Every member of the band plays at least two instruments.
3. No instrument is played by more than two band members.

From these axioms we could prove the theorem, "At least three instruments are played by the band."

In 1900, at the Second International Congress of Mathematicians held in Paris, German mathematician David Hilbert presented his famous list of twenty-three challenges, in the form of unsolved problems, to the international mathematical community. Hilbert believed these problems encompassed the most important issues facing twentieth-century mathematicians. To date, over half of these problems have been solved. Hilbert's sixth problem begins as follows [Yandell 02, pp. 159–160]:

> The investigations on the foundation of geometry suggest the problem: To treat in the same manner, by means of axioms, those physical sciences in which mathematics plays an important part; in the first rank are the theory of probabilities and mechanics.

Kolmogorov solved half of this problem by axiomatizing probability in 1933. Axiomatizing mechanics (physics) has yet to be accomplished. (Kolmogorov may also have solved Hilbert's thirteenth problem, which involves the solutions to algebraic equations. Since the problem is believed to have more than one interpretation, some believe that, at best, he made a significant contribution but that the problem remains unsolved.) But, if axiomatization is to be thought of as a foundation from which theorems are derived, then why didn't it happen earlier? Why wasn't an axiomatization of probability developed by Huygens, Pascal, Fermat, or Bernoulli? It didn't because prior to the twentieth century the mathematics required for such an axiomatization, *measure theory*, had not been sufficiently developed. In simplest terms, the *measure* of a set of points is the size of the point set. The measure of a single point is 0. The measure of the set of points on the real number line between 0 and 1 is 1 because the length of this line segment is 1. The measure of the interior of a 1×1 square is 1 because the area of the square is 1. With more complex point sets, things get a bit more interesting. For example,

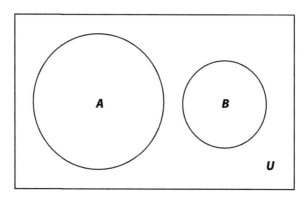

Figure 1.3. Connection between measure and probability.

the measure of all points on the real line between 0 and 1 representing *rational* numbers is 0 [Wapner 05, pp. 126–127]. (A rational number is any number that can be expressed as a ratio of two integers: 2/3, 4(4 = 4/1), 0.7(0.7 = 7/10), etc. Numbers that cannot be expressed that way, such as $\sqrt{2}$ and π, are *irrational*.) It follows that the measure of the point set on the same line segment corresponding to irrational numbers is 1. Perhaps even more curious is the existence of point sets that have no measure [Wapner 05, pp. 132–133].

Kolmogorov made the association between *measure* and *probability*. Figure 1.3 shows a *Venn diagram* representing point sets. Point set U represents all points inside the rectangle. The measure (area) of point set A, the interior of the circle on the left, is twice that of point set B, the interior of the circle on the right. Think of U as representing all possible events, or outcomes, associated with some process. Then A and B could represent specific outcomes, or collections of outcomes. The measure of A being twice that of B reflects the fact that A is two times more likely to occur than B. The point sets A and B are shown as disjoint (non-overlapping) to indicate that, as events, A and B are mutually exclusive; they cannot occur simultaneously.

The three Kolmogorov axioms of probability follow.

Axiom 1

The probability of an event is a nonnegative real number.

Axiom 2

Of all possible events, the probability that some event will occur is 1.

Axiom 3

If A_1, A_2, A_3, ... are pairwise mutually exclusive (non-overlapping) events, then the probability of either of these occurring is the sum of the probabilities of each.

Applying the axioms to Figure 1.3 and interpreting measure as area, the three axioms can be stated as follows.

Axiom 1

An area is a nonnegative real number.

Axiom 2

The area of the entire rectangular region U is taken to be 1.

Axiom 3

The area of the regions enclosed by the two non-overlapping circles is the sum of the two circular areas A and B.

It is from this simple and succinct set of three axioms that the rules of probability are derived. These rules, and an introduction to mathematical expectation, are given in the next chapter.

The ABCs of E

Perhaps the most important idea in making rational decisions in the face of uncertainty (a common task in gambling and many games) is the concept of mathematical expectation.

—Edward Packel

M athematical expectation is simple to define and its calculation involves nothing more than products and sums. The first half of this chapter defines mathematical probability and includes some basic rules; then it's on to mathematical expectation.

In the context of this book, an *experiment* is a process that leads to measurable *outcomes*. An *event* is any collection of outcomes associated with the experiment. A *simple event* consists of a single outcome; a *compound event* consists of more than one outcome. When a card is drawn from a standard deck of 52 playing cards, possible outcomes include drawing the ace of hearts, the seven of diamonds, etc. The event of drawing any black card is a compound event because it is associated with 26 outcomes, corresponding to the 26 black cards of the deck.

The *sample space S* is the set of all simple events (outcomes) associated with an experiment. The sample space associated with drawing a single card from a standard deck consists of 52 simple events, or *sample points*. The number of sample points in the sample space is significant, and we denote it by $n(S)$; it's called the *cardinality* of S and represents the number of outcomes associated with the experiment. For a randomly selected card from a standard deck, $n(S) = 52$. For the single toss of a fair coin, $S = \{H,T\}$ and $n(S) = 2$. If we take into account all events (simple and compound), associated with an experiment, then the total number of events far exceeds the number of simple events alone. For example, when a single die is rolled once, the sample space is given by $S = \{1,2,3,4,5,6\}$ and $n(S) = 6$. But there are other (compound) events that could be considered. There's the event of rolling an even number and we could think of this as {2,4,6}, a subset of *S*. The event of rolling a number greater than 4 is described by the subset {5,6}. It follows that the total number of events (simple and compound) associated with an experiment is equal to the total number of subsets of the sample space. This set of all subsets is called the *power set* of *S* and its cardinality is given by $2^{n(S)}$ [Aczel 00, pp. 149–151].

For example, for the toss of a fair coin, the sample space is given by $S = \{H,T\}$. There are two simple events; however, there are $2^2 = 4$ total events associated with this single toss. In addition to tossing heads or tails, there is the event of tossing neither heads nor tails (an impossible event, yet an event nonetheless) and the event of tossing either heads or tails (which is certain to occur).

When a single card is randomly drawn from a standard deck of 52 cards, there are 2^{52} possible events if we include compound events such as drawing a queen, drawing a club, drawing a black card, etc. If each such event were listed, one per line, in books such as this, over 500 billion such books would be required to list all members of the power set. By any measure, this far exceeds the total number of books in existence today!

The Definition of Probability

The *probability* of an event, be it simple or compound, is defined as its theoretical relative frequency of occurrence when the experiment is performed multiple times. It measures the event's likelihood. Toss a fair coin a large number of times and we would expect it to come up heads roughly 50 percent of the time. So we say the probability of tossing heads on a single toss of a fair coin is 1/2. If we roll a single die repeatedly, we would expect to roll a 1 (or any of the five other possibilities) one sixth of the time; therefore, the probability of rolling any given number on a single roll of the die is 1/6. These are examples of *true* or *theoretical probabilities* and are the most accurate form of probability. Unfortunately, such ideal forms of probability are not that easy to come by because in reality no toss of a coin or roll of a die is perfectly random and no real coin or die is perfectly shaped. All coins are biased, perhaps insignificantly so, due to the fact that the design is not the same on both sides. And no die is a perfect cube. True probabilities are based on perfection that may not exist in physical reality. In the absence of physical perfection, we may be required to use a statistical estimate of the theoretical probability, known as the *experimental* or *empirical probability*.

For example, calculating the true probability of a tack landing point up after being tossed would be virtually impossible due to its peculiar shape (see Figure 2.1). A practical way to estimate the true probability would be to toss the tack a large number of times and record the frequency of point-up landings. Equivalently, we could toss a large number of identically shaped tacks once and record the number of tacks that land point up.

Figure 2.1. Up or down?

If, say, 1,000 such tacks are tossed and 160 land point up, we could invoke Bernoulli's law of large numbers and assume the true probability is approximately equal to the experimental probability of 160/1000 = 4/25 or 16 percent. A more reliable estimate could be achieved by tossing a larger number of tacks and recording the results.

In practice empirical probabilities are used as estimates of true probabilities because the true (theoretical) probabilities are not generally accessible. The probability a patient will experience a particular side effect when taking a certain prescription medication is based on experimental data from clinical trials. Insurance companies use empirical probabilities as measures of risk for theft, accidents, injuries, and death.

In some cases, when no experimental data is available, we may be forced to use a *subjective probability*, or hunch, as our best estimate of the true probability. What is the probability a previously untried medical procedure will be successful? Will a new business venture succeed or fail? Will the couple's first date be followed by a second? Will they marry and live happily ever after?

Throughout this book we generally assume all given probabilities are theoretical in nature and therefore perfectly reflect the likelihood of occurrence of the event in question.

Denoting the probability of an event A as p (sometimes $p(A)$ or p_A), it's clear that $0 \le p \le 1$. (This can also be derived from the Kolmogorov axioms.) Events having probabilities on the high end (near 1) are highly likely to occur. Events with probabilities on the low end (near 0) are unlikely. An event that is as likely to occur as not (such as tossing heads on a single toss of a fair coin) has a probability of 1/2 or 50 percent. If an event is certain to occur, its probability is 1 or 100 percent. If an event is impossible, its probability is 0 (0 percent). Interestingly, the converse of each of the last two statements is false. That is, if the probability of an event is 100 percent, this does not, in and of itself, mean that the event is certain to occur. Similarly, if the probability of an event is 0, it need not be impossible. Examples are given later in this chapter.

But, if an experiment is performed only once, can we still speak of a relative frequency of occurrence? And what exactly does this mean? The relative frequency of occurrence definition describes a set of outcomes and doesn't describe the exact nature of any single coin toss or roll of a die. After all, the coin either comes up heads, or it doesn't.

Years ago I underwent a minor surgical procedure and returned to see the surgeon for follow up. During my return visit I asked the doctor what my chances were of developing complications. He replied, "I don't know the exact probability. But in your case it's either 100 percent or 0, so you might as well not worry about it." I understood what he meant; but, his answer lacked something.

Probability may not describe, literally, a single event; but, it is indeed a measure of the event's likelihood. If, for example, the relative frequency of occurrence of heads for a biased coin is 99 percent, then, on average, the coin will come up heads 99 times for every 100 times it is tossed. Since this is the case, it would be highly likely that the coin will come up heads for any given toss.

The Laws of Probability

The laws of probability are derivable from Kolmogorov's three axioms and are given here without derivation.

For an experiment with a sample space having equally likely sample points (simple events), known as *equiprobable*, the probability of each sample point is given by $1/n(S)$. A basic notion follows.

A Fundamental Notion of Equiprobable Sample Spaces

For an experiment with an equiprobable sample space S, the probability of an event A is given by $p(A) = \frac{n(A)}{n(S)}$.

For example, the probability of randomly selecting the ace of spades from a standard deck of 52 cards is 1/52 as there is only one ace of spades that may be chosen from the 52 equiprobable cards. The probability of randomly drawing an ace (any ace) is 4/52 = 1/13 as there are four aces in a standard deck.

When a standard pair of six-sided dice is rolled, the sample space consists of 36 outcomes. For each of the six possible outcomes of one die there are six possible outcomes for the other. If each simple event is portrayed in the form (*outcome of the first die, outcome of the second die*), then the sample space is given by

$$S = \begin{cases} (1,1), & (1,2), & (1,3), & (1,4), & (1,5), & (1,6), \\ (2,1), & (2,2), & (2,3), & (2,4), & (2,5), & (2,6), \\ (3,1), & (3,2), & (3,3), & (3,4), & (3,5), & (3,6), \\ (4,1), & (4,2), & (4,3), & (4,4), & (4,5), & (4,6), \\ (5,1), & (5,2), & (5,3), & (5,4), & (5,5), & (5,6), \\ (6,1), & (6,2), & (6,3), & (6,4), & (6,5), & (6,6) \end{cases}.$$

Sum	Probability
2	1/36
3	2/36 = 1/18
4	3/36 = 1/12
5	4/36 = 1/9
6	5/36
7	6/36 = 1/6
8	5/36
9	4/36 = 1/9
10	3/36 = 1/12
11	2/36 = 1/18
12	1/36

Table 2.1. Sums and probabilities for a standard pair of dice.

Rolling a sum of 2 can be done in only one way, as (1,1); therefore, the probability of rolling a sum of 2 is 1/36. Rolling a sum of 3 is twice as likely because rolling a 3 can be done in two ways: (1,2) and (2,1). The most likely sum to roll is 7. The six ways of doing so are depicted in *S* along the diagonal running from the lower left to the upper right. The probability of rolling a 7 is therefore 6/36 = 1/6. The probabilities of rolling each of the possible 11 sums are given in Table 2.1.

With these probabilities in mind, the law of large numbers can be illustrated. The first two histograms (vertical bar graphs) of Figure 2.2 give the relative frequencies of sums from an experiment in which a pair of dice was rolled 20 and 100 times. (These are the results from one experiment; your own results might be very different.) The third histogram gives the true probabilities (theoretical relative frequencies) of the sums. The law of large numbers is visually confirmed by noting how the shapes of the histograms approach the third histogram of true probabilities with the larger number of rolls.

The formula for $p(A)$ given above only applies in situations where the sample space is equiprobable. For example, consider the toss of a relatively thick coin, such as a US nickel (5¢), where the sample space includes the possibility of the coin landing on its edge. Landing on its edge corresponds to one of the three outcomes; yet, it would be ludicrous to suggest the probability of this occurring is 1/3. Such faulty reasoning would suggest your probability of living to see tomorrow is 1/2 because there are two possibilities and survival corresponds to one of the two. Is there life on Mars? Would you argue there is a 50 percent chance of this because there are two possibilities? Hopefully not!

The *complement* of event *A*, denoted as $\bar{A}$, is the set of all sample points in the sample space not belonging to *A*. In other words, $\bar{A}$ is the event that *A* does not occur. If a single die is rolled and $A = \{2,4,6\}$, then $\bar{A} = \{1,3,5\}$.

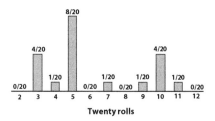

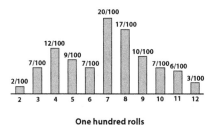

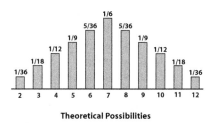

Figure 2.2. A pair of dice and the law of large numbers.

The Complement Rule

For any two complementary events A and Ā, p(Ā) = 1 − p(A).

So, if there is a 10 percent chance of rain tomorrow, we equivalently conclude there is a 90 percent chance it will not rain.

Two events, *A* and *B*, are *mutually exclusive* (or *disjoint*) if they cannot occur at the same time. When a single card is drawn from a standard deck of 52 cards, drawing a queen and drawing a king are mutually exclusive. These events are also described as disjoint because as sets they have nothing in common. The first event is *Q* = {Q♠, Q♥, Q♦, Q♣} and the second is *K* = {K♠, K♥, K♦, K♣}; there is no overlap. The compound event "queen or king" would be the event of drawing either a queen or a king on a single draw from the deck. The probability of drawing a queen or a king is then

$$p(Q \text{ or } K) = \frac{4+4}{52} = \frac{4}{52} + \frac{4}{52} = p(Q) + p(K) = \frac{2}{13}.$$

We simply add the two individual probabilities to get the probability of the desired compound event.

The Addition Rule for Mutually Exclusive Events

For any two mutually exclusive events A and B, p(A or B) = p(A) + p(B).

The addition rule is one of the Kolmogorov axioms and may be generalized for pairs of events that need not be mutually exclusive. Using the standard deck as above, let R denote the event of drawing any red card and let Q denote the probability of drawing any queen. Despite the fact there are 26 red cards and 4 queens, it would be incorrect to conclude there are 26 + 4 = 30 ways to draw a card of either type. The error occurs because two red cards are being counted twice—once as red cards and once again as queens. The actual number of cards in a deck to be considered as red cards or queens is given by

$$n(R \text{ or } Q) = n(R) + n(Q) - n(R \text{ and } Q) = 26 + 4 - 2 = 28$$

in which case the probability of drawing any red card or any queen on a single draw is 28/52 = 7/13. This suggests the following general rule for finding the compound probability of either of any two events occurring.

The General Addition Rule

For any two events A and B, p(A or B) = p(A) + p(B) − p(A and B).

In the case that events A and B are mutually exclusive, then $p(A \text{ and } B) = 0$ and the formula is effectively the same as that for mutually exclusive events.

Sometimes additional information presents itself in the form of a preexisting condition (event) that effectively shrinks the size of the experiment's sample space. For example, the probability of a randomly chosen US born male individual being diagnosed with prostate cancer at some time in their life is approximately 18 percent. But if it is determined that a close family (father, brother) history of this form of cancer exists, then the sample space shrinks to a smaller group of higher-risk individuals and the lifetime risk of this diagnosis jumps to 45 percent. The lifetime risk of melanoma of the skin for all US individuals is approximately 3 percent. But if a randomly chosen individual is black, the risk falls dramatically to 0.1 percent because the reduced sample space of black individuals are far less likely to develop skin cancer than the general US population.

So, additional information may increase or decrease the likelihood of an event; these are known as *conditional probabilities*. The "fortunately-unfortunately" tale below shows an increased and decreased likelihood of a happy ending as additional information becomes available:

There was a man flying up in the sky.
Fortunately he was in an airplane.
Unfortunately he fell out of the plane.
Fortunately he had on a parachute.
Unfortunately the parachute didn't open.
Fortunately there was a haystack below.
Unfortunately there was a pitchfork in the haystack.
Fortunately he missed the pitchfork.
Unfortunately he missed the haystack!

In order to establish a formula for calculating conditional probabilities, we consider once again the simple act of randomly drawing one card from a standard deck. The probability of drawing the king of hearts is $p(K\heartsuit) = 1/52$. We draw the card and before we have a chance to look at it, we are told reliably that the card is a king. Now the sample space of possible cards shrinks to a cardinality of 4—{$K\spadesuit$, $K\heartsuit$, $K\diamondsuit$, $K\clubsuit$}—and the probability of drawing the king of hearts, given that the card is a king, is $1/4$. If we divide the numerator and denominator by 52, the cardinality of the original sample space (which is the same as multiplying by one), the conditional probability of drawing the king of hearts, given that the card is a king, can be given as $\frac{1/52}{4/52}$, which represents the probability of drawing a king and a heart, divided by the probability of drawing any king. This suggests the following formula for computing conditional probabilities.

The Conditional Probability Rule

For any two events A and B, the conditional probability of A, given the occurrence of B, is denoted as p(A|B) and may be calculated as

$$p(A|B) = \frac{p(A \text{ and } B)}{p(B)}.$$

Two comments and a short discussion about this formula are in order. First, the derivation of this rule assumes the sample space to be equiprobable. It can be shown that this formula applies as well for general sample spaces which need not be equiprobable. Second, we need not use the formula to compute simple, conditional probabilities. Visualizing the reduced sample space may be all that is required to arrive at the desired probability.

A distinction between $p(A|B)$ and $p(B|A)$ is needed. Note that

$$p(A|B) = \frac{p(A \text{ and } B)}{p(B)} \neq \frac{p(A \text{ and } B)}{p(A)} = p(B|A).$$

In plain English, given that your pants are on fire, it is highly likely you are distressed. But given that you are distressed, it is not likely that your pants are on fire.

This silly example clearly differentiates the two probabilities. But, in more involved and realistic scenarios, the confusion becomes problematic. The global pharmaceutical company Pfizer Incorporated, in promoting its cholesterol medication Lipitor, makes the following claim:

> High cholesterol is a risk factor for heart attack and stroke. In fact, about every 34 seconds in the US, someone has a heart attack, and 80 percent of those people have high cholesterol.

The claim may be true; however, it would be false to assume the claim suggests there is an 80 percent chance of someone with high cholesterol having a heart attack. (In the US approximately 80 percent of shark attack victims are right handed. It would be clearly incorrect to assume there is an 80 percent chance of a right-handed individual being attacked by a shark.) The inability to make the distinction is known as *confusion of the inverse* and may occur with serious, if not life-threatening, consequences. Two examples follow.

A virus is known to infect 1 percent of a population, and a relatively accurate test is used to determine if an individual is infected. The accuracy of the test is given in terms of its *sensitivity* (the probability the test will report positive, given that the individual is infected) and its *specificity* (the probability the test will report negative given no presence of the virus). Let's assume both measures are 90 percent, yielding an overall accuracy of 90 percent as well. Using V to denote the event of the virus being present in an individual, we have the test's accuracy given by the following two measures:

sensitivity $= p(+|V) = .90$ and specificity $= p(-|\overline{V}) = .90$.

So, of all tests given, 90 percent are reported accurately, be the results positive or negative. Ten percent are incorrect, coming back as either false positive or false negative results. If 1,000 individuals are randomly selected from the population and tested for the virus, we would expect theoretical results as depicted in Table 2.2.

Assume that you are one of those tested and the test results are positive. Should you and your physician be concerned? Of course! But how likely is it that

Test Results

Condition of Patient	Positive	Negative	Total
Virus	9	1	10
No virus	99	891	990
Total	108	892	1,000

Table 2.2. Medical diagnostic testing.

you have the virus? With the understanding that the test is 90 percent accurate, you might assume there is a 90 percent chance you are infected. But, this is far from correct. It is true that $p(+|V)$ = .90; however, we must not confuse this with the desired probability of $p(V|+)$, the *predictive value of a positive result*. Using Table 2.2 we see that $p(V|+) = \frac{9}{108} \approx .08$ or 8 percent, far less than the 90 percent sensitivity. This discrepancy is significant with respect to both diagnosis and treatment. In all likelihood you are not infected and the test result was a false positive. As Table 2.2 indicates, roughly 92 percent (99/108) of positive results are false positives.

Should you and your physician ignore the test results? Definitely not! Despite the fact the probability of your being infected is only 8 percent, you are now eight times more likely to be infected knowing that you have tested positive. Perhaps a retest or other diagnostic tool is in order. Or, if treatment is safe and inexpensive, it might be best to be treated for the virus, even if you are not infected. But wouldn't it be best to err on the side of caution and treat in all cases? This would depend on the exact nature of the malady being diagnosed and the nature of the treatment. In some cases treatment for a nonexisting condition may be risky (surgery, drugs with debilitating side effects), be expensive, and put the patient through unnecessary physical and psychological stress. Diagnosis and treatment should follow careful consideration of the nature of the virus, the cost of treatment, and the distinction between the sensitivity of the test and the predictive value of a positive result.

With respect to diagnosis and treatment, a test's predictive value (with respect to a positive result) is more significant than the test's overall accuracy. Consider what would happen if a medical laboratory used the highly unethical and illegal practice of declaring all test results negative, without actually testing the samples obtained. (A few random positive results might be included so as to make the results appear legitimate.) For the previously described virus, the overall accuracy of the test would increase from 90 percent (legitimate testing) to 99 percent. Ethics aside, wouldn't patients and physicians prefer 99 percent accuracy to 90 percent accuracy? And, wouldn't this practice be more cost effective, since no actual testing would be required? The answer to both questions is "no". If positive results are only reported randomly, then the predictive value of a positive re-sult drops to 1 percent, making the test almost worthless for detecting the virus.

The judicial equivalent of a medical diagnostic test is a court trial, where evidence is presented and a determination is made as to whether or not the de-fendant is guilty of some crime. A conviction would be analogous to a positive medical test result and an acquittal would be similar to a negative test result. Over the past twenty years, forensic evidence in the form of DNA profiling has become a powerful tool used by both the prosecution and the defense in paternity cases and criminal trials. A misunderstanding of the statistics associated with DNA profiling in the form of confusion of the inverse can lead to false convictions and other errors in judgment.

For this discussion, assume semen obtained from a rape victim undergoes DNA analysis and has a profile (fingerprint) that occurs once in every one mil-lion individuals. That is, the probability of an innocent person having this same profile is .000001. If an innocent person is accused of this crime, the conditional probability of a match is

$$p(\text{DNA match}|\text{innocent}) = .000001.$$

An individual is arrested and accused of committing the rape. A DNA sample is taken and the profile matches that of the collected evidence. To be sure, the evi-dence is damning. Based on the match alone, how likely is it that the defendant committed the rape?

The *prosecutor's fallacy* occurs in assuming that

$$p(\text{innocent}|\text{DNA match}) = .000001.$$

Once again we see the confusion of the inverse. The media often incorrectly report such incorrect statements as, "Based on DNA evidence, there is a .000001 probabil-ity that the defendant is innocent" or "there is a .999999 likelihood that the defen-dant committed the rape." Jurors may believe this to be the case, whether or not the prosecution presents the evidence in this false way. A crafty, but perhaps unethi-cal prosecutor could present the evidence correctly yet word it in such a way that jurors are likely to misinterpret the evidence in a way that favors the prosecution.

Of course, the defense can distort the truth as well. If the rape occurs in a city of eight million people, then approximately eight individuals are expected to pos-sess the DNA profile associated with the semen. This suggests that the defendant may be considered one of eight possible individuals who may have committed the crime. In such a case, his probability of guilt may be considered to be 1/8, far from the level of certainty needed for a conviction.

In actuality, it's difficult to argue one way or the other, without more evi-dence. As is the case with medical diagnostic testing, additional evidence is re-quired and great care should be taken in the absence of such.

We can also consider combining probabilities in a different way. The addition rule allows us to compute probabilities of the form $p(A$ or $B)$. Probabilities of the form $p(A$ and $B)$ can be calculated using an equivalent form of the conditional probability rule. Multiplying both sides of $p(B|A) = \frac{p(A \text{ and } B)}{p(A)}$ by $p(A)$ yields the following formula.

The General Multiplication Rule

For any two events A and B, the probability of A and B, denoted as p(A and B), may be calculated as p(A and B) = p(A)p(B|A).

Two events are considered *independent* if the occurrence of one has no influence on the likelihood of occurrence of the other. If events A and B are independent, then the probability of A given B is simply equal to the probability of A, with condition B ignored. That is $p(A|B) = p(A)$. Similarly, $p(B|A) = p(B)$. This leads to a multiplication rule for the special case of events A and B being independent.

The Multiplication Rule for Independent Events

For any two independent events A and B, p(A and B) – p(A)p(B).

To clarify, consider four cards drawn from a standard deck of 52 cards, one at a time, with each card noted and replaced before the next draw. We wish to compute the probability of all four cards being aces. That is, the first card is to be an ace, *and* the second card is to be an ace, *and* so on. The four events forming the compound event are independent because the card is being replaced each time and the deck will have no recall of previous draws. So, the probability is given by

$$\frac{4}{52} \times \frac{4}{52} \times \frac{4}{52} \times \frac{4}{52} = \frac{1}{28{,}561} \approx .000035.$$

On the other hand, if the four cards are dealt from the deck without replacement, then the deck *remembers* previous draws and the sequential factors must take into account the condition that aces have been dealt and removed from the deck. Under these conditions, the probability of dealing four aces (without replacement) from a standard deck is given by

Number of aces left after each draw

$$\frac{4}{52} \times \frac{3}{51} \times \frac{2}{50} \times \frac{1}{49} = \frac{1}{270{,}725} \approx .0000037,$$

Cards left in deck after each draw

roughly one tenth as likely as with replacement.

Earlier in this chapter it was stated that if the probability of an event is 0, it need not be impossible and if the probability of an event is 1 it need not be certain. Using the multiplication rule for independent events, we show why this is so. Imagine a fair coin is to be tossed indefinitely many times, until heads first appears. What is the probability of the coin never coming up heads, no matter how many times it is tossed? For this to occur, it would have to come up tails and tails and tails ... infinitely many times. The probability of this is given by

$$\frac{1}{2} \times \frac{1}{2} \times \frac{1}{2} \times \frac{1}{2} \times \ldots,$$

which approaches 0 as the number of tosses (factors) increases. The probability of heads never occurring is 0. Yet, there is no physical reason for it to be impossible. In principle, tales could occur infinitely many times. Mathematically the event is described as being *effectively impossible*, occurring with probability 0, though in actuality it could occur. Equivalently the probability that the coin will eventually come up heads is 1, despite the fact this need not occur. The event that the coin eventually comes up heads is said to be *almost certain* and occurs with probability 1, though it need not occur at all!

Binomial Probabilities

A common type of experiment involves a sequence of n identical trials, each of which has two possible outcomes—*success* or *failure*. In a *binomial* or *Bernoulli trials* experiment we are interested in the probability of exactly x successes, ignoring the order in which the successes appear. The terms *success* and *failure* need not be taken literally; they are used primarily to distinguish the two possible outcomes associated with each of the n trials. For example, a couple chooses to have four children, and we want to know the probability that exactly three of the four will be girls. Order is irrelevant and we arbitrarily designate *girl* as a success and *boy* as a failure.

The precise conditions for a binomial trials experiment are as follows:

1. Each of the n identical trials has two possible outcomes—*success* or *failure*. One and only one of these two outcomes must occur on each of the trials.
2. The outcome of any one trial has no effect on the outcome of any other trial. That is, the outcomes are independent.
3. The probability of success, denoted by p, remains constant for each trial. Consequently, the probability of failure, denoted by q, where $q = 1 - p$, also remains constant.

For any specified set of x trials, the probability of all successes is given by p^x and the probability of the remaining $n - x$ trials all being failures is q^{n-x}. So, for the given trials being successes and failures as specified, the probability is $p^x q^{n-x}$. But the x successes need not occur in any particular order. For example, the couple wanting four children can have two girls (and two boys) in six possible ways— *ggbb, bggb, bbgg, gbgb, bgbg,* and *gbbg.* In general, the number of ways x successes (and $n - x$ failures) can be specified from a set of n trials is given by

$$\frac{n!}{x!(n-x)!}, \quad \text{where } n! = n \text{ factorial} = n(n-1)(n-2)\ldots 2 \cdot 1.$$

Using the previously established rules, we are led to the formula that gives the probability of obtaining exactly x successes in n binomial trials.

The Binomial Probability Formula

The probability of obtaining x successes in n binomial trials is given by

$$P(x) = \frac{n!}{x!(n-x)!} p^x q^{n-x}.$$

The probability that exactly two of four children are girls is $\frac{4!}{2!2!}(1/2)^2(1/2)^2$ = 3/8. The probability of there being exactly one girl (and three boys) is $\frac{4!}{1!3!}(1/2)^1$ $(1/2)^3 = 1/4$, which is also equal to the probability of there being exactly one boy (and three girls). The probability of there being three of one sex and one of the other (three girls and one boy or one girl and three boys) is therefore $1/4 + 1/4 = 1/2$. Some readers may be surprised that it is more likely for the four children to be three of one sex and one of the other, rather than evenly split as two of one and two of the other.

The Definition of Expected Value

For an experiment having numerical outcomes, the *expected value* or *mathematical expectation* of the experiment is the long-term average value of the outcome. To arrive at a precise definition and a formula for calculating the expected value, let's consider rolling a fair die and winning, in dollars, the number rolled. That is, on each roll of the die the player will win $1, $2, $3, $4, $5, or $6, depending on the roll's outcome. If the die is rolled six times, we would expect to roll each number, on average, once. This is due to the die's symmetry. It may be unlikely for each number to come up exactly once in six rolls; however, it is more likely for this to occur than any other specified outcome because the simple events are equiprobable due to symmetry. So, the average win per roll of the die would theoretically be

$$\frac{1 \times \$1 + 1 \times \$2 + 1 \times \$3 + 1 \times \$4 + 1 \times \$5 + 1 \times \$6}{6} = \$3.50.$$

Obviously no single roll of the die yields a win of $3.50 as no face on the die shows 3.5 pips. But, if the die were rolled a large number of times, the player should expect to win, on average $3.50 per roll. After 100 such rolls, the player is expected to have won approximately $350.

Another way to interpret this is in terms of what the player should be willing to pay, up front, in order to play this game. Rationally, the player should be willing to pay anything less than $3.50 because, in doing so, the player is expected, in the long run, to come out ahead. For example, if the player were charged $3.00 to roll the die and then win the amount rolled in dollars, the player's expectation would be $3.50 − $3.00 = $.50 per roll and, on average, would net $.50 per roll. On the other hand, being charged $4.00 to roll the die would give the player a negative expectation of $3.50 − $4.00 = −$.50, costing the player, on average $.50 per roll of the die.

The expression used in the calculation for the six sided die can be rewritten as

$$\frac{1}{6}(\$1)+\frac{1}{6}(\$2)+\frac{1}{6}(\$3)+\frac{1}{6}(\$4)+\frac{1}{6}(\$5)+\frac{1}{6}(\$6),$$

which is of the form $p_1 x_1 + p_2 x_2 + \dots + p_n x_n$, where p_i is the probability associated with the outcome (payoff) x_i. A compressed way to indicate the sum is to write it in sigma notation as $\sum_{i=1}^{n} p_i x_i$, indicating the sum of terms of the form $p_i x_i$ as i takes on values from 1 though n. For example,

$$\sum_{i=1}^{6} p_i x_i = p_1 x_1 + p_2 x_2 + p_3 x_3 + p_4 x_4 + p_5 x_5 + p_6 x_6.$$

In mathematics, such a sum of terms in a sequence is called a *series*. Referring to a numerical outcome as a *random variable*, we have the following formula.

The Expected Value of a Random Variable

The expected value of a random variable is equal to the sum of all possible values of the variable, each multiplied by its respective probability. Symbolically,

$$E = \sum_{i=1}^{n} p_i x_i \quad or \quad E = \sum_{i=1}^{n} x_i p_i.$$

In the card game blackjack (twenty-one) the ace is worth one point or 11 points, whichever is most advantageous to the player. The face cards (jacks, queens, and kings) are all worth ten points each. There are 12 such face cards in the deck. All other cards are worth, in points, their numeric value. Considering an ace to be worth one point, if we randomly select a single card from the deck, its expected point value is

$$E = \frac{4}{52}(1) + \frac{4}{52}(2) + \ldots + \frac{16}{52}(10) = \frac{85}{13} \approx 6.5.$$

If the ace is considered to be worth 11 points, then the expected point value of a single card is

$$E = \frac{4}{52}(2) + \frac{4}{52}(3) + \ldots + \frac{16}{52}(10) + \frac{4}{52}(11) = \frac{95}{13} \approx 7.3.$$

The expected value formula may be used to calculate the expected value of a *binomial* random variable; however, there's an easier way to do so. The expected number of girls in a family of four children can be found by evaluating

$$E = \sum_{i=0}^{4} \frac{4!}{i!(4-i)!}\left(\frac{1}{2}\right)^i \left(\frac{1}{2}\right)^{4-i} = 2,$$

but the answer is intuitive because the theoretical relative frequency of a girl is 1/2 and there are four trials. In general, for n binomial trials, the expected number of successes is simply $E = np$.

A final point must be made with respect to the applicability of mathematical expectation to short-term or even single-play scenarios. If there is no long term, then would it make sense to think in terms of a theoretical long-term average for a single play of the game? In some sense it would. For a single roll of a fair die, we do not *literally* expect to roll 3.5, despite this being the mathematical expectation of such a roll. We use the expectation in more of a comparative sense, making rational decisions regarding which games to play and how much should be wagered by comparing expectations and wagers of the various options before us. In writing about social conflict and social physics, professor of psychology and mathematics Anatol Rapoport writes [Rapaport 74, pp. 161–163],

> But what if there is no long run? Where there is no long run (as in one-shot games), the expected utility idea must be considered simply as an extrapolation of a long run policy. ... It seems the best one can do. ... The same kinds of rules are then also extended to situations where there is no "long run", because one needs some operating rule. Without operating rules (and these must always be to a certain extent arbitrary) no decisions at all are possible. ... The long run expectations mean nothing in the individual case, but the rules at least make rationalizable decisions possible. This is important in situations where no decision may be worse than a bad decision.

Utility

Throughout most of this book, the expressions *payoff, value,* and *worth* are to be considered synonymous. An individual receiving a payoff of $100 receives something valued at $100, which is worth $100, and so on. But, in some cases, a careful distinction should be made between a numerical or monetary payoff (value) and the payoff's intrinsic worth to the individual (the individual's satisfaction), known as *utility*. In some cases, the utility of a payoff is not proportional to its monetary value.

Do you value $10 twice as much as $5? Probably you do. Now imagine waking to the unbelievably good news that you're the winner of a $5 million state lottery. Or is it $10 million? What's the difference? Rich is rich! The sums are so large, there is probably little difference to you, one way or the other. Do you value $10 million twice as much as $5 million? Probably not.

Figure 2.3 illustrates how utility often measures satisfaction in a diminishing returns sort of way.

It is assumed that rational individuals make choices so as to maximize the expected value of the return associated with their choice. But, in situations where value and utility differ, the rational individual will presumably play so as to maximize the expected utility rather than the expected value. As an example, an adult playing a game against a child may prefer playing to lose, getting more satisfaction out of losing than winning points or dollars. Rather than playing to maximize accumulated points, the adult maximizes the intrinsic worth of the game, gaining a higher level of satisfaction by losing rather than winning. Throughout this book we treat *payoff, value,* and *utility* as synonymous, unless their distinction is significant as in the previous examples. An amusing example follows. We see that as additional information becomes available, expectations can flip-flop in much the same way as conditional probabilities change with additional information. Ultimately, it is the notion of utility that settles the matter.

Figure 2.3. Diminishing returns.

A professional baseball team is scheduled to play 162 games in any given season. Let's assume your favorite team will average about five runs per game (win or lose) so that a list of its scores for the 162 games might appear something like 6, 2, 5, 7, ... Before the season begins, you are asked to choose one of two possible payoffs, which you will receive after the last game of the season is played. You may elect to receive the sum of all your team's scores in dollars—$(6 + 2 + 5 + 7 + ...)—or you may chose to receive the product of all scores in dollars—$(6 × 2 × 5 × 7 × ...). You must now decide which of the two options will yield the greater expectation. At first glance, almost everyone chooses the "×" option because the product appears to have a much larger expected value than the "+" option.

But then it should occur to you that your team will almost certainly be shutout and score 0 runs at least once during the season. This being so, it is highly likely that the product will be 0, whereas the sum is expected to be about 162 × ($5) = $810 With this in consideration, you will most likely change your selection to the "+" option.

But maybe we should reconsider yet again. In the unlikely event your team is never shutout, the "×" payoff will be astronomically large. Shouldn't a proper analysis compare the two expectations associated with the two options? Perhaps the large potential payoff associated with the "×" option will make up for its unlikelihood. Making the realistic assumption that a team will be shutout with probability .05 and denoting the two expected payoffs as E_+ and E_x, we compare the two expectations:

$$E_+ = \$(5 + 5 + 5 + ...5) = 162(\$5) = \$810, \quad E_x \approx (.95)^{162}(5^{162}) \approx 4.2 \times 10^{109}.$$

It's not even close! The "×" option yields an astronomically greater expectation than the "+" option, and it now appears that the product is indeed the better of the two choices. Our initial choice of this option may have been correct, despite the fact that it was made without considering the significant likelihood of a shutout. The initial choice of "×" was the right choice, though based on incomplete information.

Unexpectedly, we're not through with this problem yet. Are you convinced the "×" option is preferable? Which option would you choose? Despite the preceding analysis, most would stick with the "+" option. Why? Maybe there's yet even more information to consider. Here's where the notion of utility is important. The potential payoff associated with the "×" option is mind-numbingly large, albeit highly unlikely. It is on the order of (1.7×10^{113}), and we tend to appreciate this in terms of its intrinsic worth, or utility, rather than its actual value. So, we compare the expected utilities of the two options, rather than the expected payoffs. To simplify the discussion we use the common (base 10) logarithm of a payoff as a measure of its utility. (The common logarithm of a positive real number is the power to which 10 must be raised to yield the number. For example, $\log 100 = 2$ because $10^2 = 100$.) The graph of the common logarithm function is concave down, reflecting the diminishing returns nature of a utility function. We now compute and compare the expected utilities associated with both options, denoting them respectively as EU_+ and EU_x:

$$EU_+ = \log(5 + 5 + 5 + ...5) = \log(162 \times 5)$$
$$= \log 810 \approx 2.9, EU_x \approx (.95)^{162}\log(5^{162}) \approx .03.$$

And now, once again, the "+" option is favorable. This final argument, in terms of utility, is the proper explanation of why the "+" option should be, and most often is, chosen. The earlier choice of this option was, once again, the right answer, but was made with incomplete information.

There are examples from all areas of mathematics where incomplete or incorrect reasoning accidentally yields a correct answer. The American Mathematical Association of Two Year Colleges (AMATYC) uses the expression "Lucky Larry" to describe the student who "solves" a problem incorrectly, yet ends up with the correct answer. Of course, Lucky Larry would earn no more credit from his instructor than a chuckle. Examples are given in this chapter's appendix.

Infinite Series: Some Sum!

Some experiments have potentially infinitely many outcomes. For example, a fair coin is tossed repeatedly until heads first appears and the number of required tosses is recorded. The value of this random variable may be any natural number with no upper limit. To compute the expected number of tosses required, we must evaluate a sum having infinitely many terms since there are infinitely many possible values of the random variable. But what exactly do we mean by a sum with infinitely many terms? Can such a sum be finite?

The sum of infinitely many terms can indeed be finite, even if all terms are positive. Consider the repeating decimal fraction expression for 1/3. We can express it as an infinite series:

$$\sum_{i=1}^{\infty} \frac{3}{10^i} = .3 + .03 + .003 + ... = .333... = \frac{1}{3}.$$

Similarly,

$$\sum_{i=1}^{\infty} \frac{6}{10^i} = .6 + .06 + .006 + ... = .666... = \frac{2}{3}.$$

(Adding the corresponding terms of both equations shows that the repeating decimal fraction .999... is exactly equal to 1, without rounding.)

Of course, not all infinite series are of finite value. The sum 1 + 1 + 1 + ... is infinitely large, increasing without bound as additional terms are added. We need a precise definition of infinite sum and a way to calculate such a sum when its value is finite.

Consider the infinite series

$$\sum_{i=1}^{\infty}\frac{1}{2^i}=\frac{1}{2}+\frac{1}{4}+\frac{1}{8}+\ldots+\frac{1}{2^n}+\ldots.$$

Create partials sums by adding only a finite number of terms. The first partial sum is simply the first term, $\frac{1}{2}$. The second partial sum is the sum of the first two terms, $\frac{1}{2}+\frac{1}{4}=\frac{3}{4}=1-\frac{1}{4}$. The third partial sum is the sum of the first three terms, $\frac{1}{2}+\frac{1}{4}+\frac{1}{8}=\frac{7}{8}=1-\frac{1}{8}$. It's easy to show (by mathematical induction) that the nth partial sum, denoted as s_n, is given by $s_n=1-\frac{1}{2^n}$. The sum of an infinite series is defined as the limit of this partial sum as n approaches infinity. Using s to denote the sum of an infinite series, we write $s=\lim_{n\to\infty}s_n$. Note that $1/2^n$ can be made arbitrarily small for sufficiently large values of n. So, for the example given,

$$s=\lim_{n\to\infty}s_n=\lim_{n\to\infty}\left(1-\frac{1}{2^n}\right)=1.$$

It is also said that the series *converges* to 1. It is only in the sense that the partial sums become arbitrarily close to 1 that we say the sum is 1. If this limit does not exist in a finite sense, then we say the series *diverges*. The infinite series 1 + 1 + 1 + ... is infinitely divergent because the nth partial sum is $s=\lim_{n\to\infty}s_n=\lim_{n\to\infty}n=\infty$. The infinite series $\sum_{i=1}^{\infty}(-1)^{i-1}=1-1+1-1+\ldots$ is a bit more interesting. The first, second, third, and subsequent partial sums are given respectively as 1, 0, 1, 0, ... The nth partial sum is given by $s_n=(1+(-1)^{n-1})/2$, which fails to converge as n approaches infinity. The infinite series has no sum, finite or otherwise. (The infinite series 1 + 1 + 1 + ... diverges *to infinity*, but 1 − 1 + 1 − 1 + ... just diverges.)

However, an incorrect analysis could yield a false conclusion. Rules that apply to finite series and convergent infinite series with positive terms do not apply to divergent series. One cannot regroup and/or rearrange the terms of a divergent series (and certain convergent series) and expect the sum (or nature of divergence) to match that of the original series. Regrouping the terms of the above divergent series can produce fallacies in the form of "false sums," two of which are given below.

Fallacy 1: The Sum is 0

$$\sum_{i=1}^{\infty}(-1)^{i-1}=1-1+1-1+\ldots=(1-1)+(1-1)+(1-1)+\ldots=0+0+0+\ldots=0$$

Fallacy 2: The Sum is 1/2

Let $s = 1 - 1 + 1 - 1 + \ldots = 1 - (1 - 1 + 1 - 1 + \ldots) = 1 - s$. It follows that $2s = 1$ or $s = 1/2$.

The *harmonic series* $\sum_{i=1}^{\infty} \frac{1}{i} = \frac{1}{1} + \frac{1}{2} + \frac{1}{3} + \cdots$ diverges to infinity, but the partial sums grow at a painfully slow rate. The sum of the first 100 terms is only slightly more than 5 and the sum of the first 1 million terms is just over 14. Summing the first 2 million terms yields a sum slightly over 15. Like Aesop's tortoise, the partial sums slowly but surely grow to exceed any specified finite positive number, no matter how large.

Showing that the harmonic series is infinitely divergent is rather simple. Note the following partial sums:

$$s_1 = 1,$$

$$s_2 = 1 + \frac{1}{2},$$

$$s_4 = 1 + \frac{1}{2} + \left(\frac{1}{3} + \frac{1}{4}\right) > 1 + \frac{1}{2} + \left(\frac{1}{4} + \frac{1}{4}\right) = 1 + \frac{2}{2},$$

$$s_8 = 1 + \frac{1}{2} + \left(\frac{1}{3} + \frac{1}{4}\right) + \left(\frac{1}{5} + \frac{1}{6} + \frac{1}{7} + \frac{1}{8}\right) > 1 + \frac{1}{2} + \left(\frac{1}{4} + \frac{1}{4}\right) + \left(\frac{1}{8} + \frac{1}{8} + \frac{1}{8} + \frac{1}{8}\right) = 1 + \frac{3}{2}.$$

In general, $s_{2^n} \geq 1 + \frac{n}{2}$, which becomes infinitely large as n approaches infinity. The partial sums grow without bound, and the harmonic series is infinitely divergent.

Appendix

Some Lucky Larry "solutions" are given below, categorized by subject. In each case the answer is correct. Not all's well that ends well!

Arithmetic

$$1^3 + 2^3 = 1 \times 3 + 2 \times 3 = 3 + 6 = 9$$

$$\left(\frac{27}{8}\right)^{2/3} = \frac{27}{8} \times \frac{2}{3} = \frac{54}{24} = \frac{9}{4}$$

$$\frac{1\cancel{6}}{\cancel{6}4} = \frac{1}{4}$$

$$\frac{1\cancel{9}}{\cancel{9}5} = \frac{1}{5}$$

$$\frac{2\cancel{6}\cancel{6}}{\cancel{6}\cancel{6}5} = \frac{2}{5}$$

Algebra

$$(x + 3)(2 - x) = 4$$

$$x + 3 = 4 \text{ or } 2 - x = 4$$

$$x = 1 \text{ or } x = -2$$

Trigonometric Limits

$$\lim_{x \to 0} \frac{\sin ax}{bx} = \left(\frac{a}{b}\right) \lim_{x \to 0} \frac{\sin x}{x} = \frac{a}{b} \cdot 1 = \frac{a}{b}$$

Beyond Description

$$\text{"T-W-}\cancel{E}\text{-}\cancel{L}\text{-}\cancel{V}\text{-}\cancel{E}\text{"} - \text{"}\cancel{E}\text{-}\cancel{L}\text{-}\cancel{E}\text{-}\cancel{V}\text{-}\cancel{E}\text{-}\cancel{N}\text{"} + \text{"O-}\cancel{N}\text{-}\cancel{E}\text{"} = \text{"T-W-O"}$$

Even broken clocks are right twice a day!

Doing the Right Thing

Life is largely a matter of expectation.
—Horace

K nowing the mathematical expectation associated with a course of action allows one to make projections based on how many times the action is performed. If an investment is expected to earn $1,000 per year and we hold the investment for ten years, then the best estimate of our total earnings after ten years would be $10,000. Predictions of this sort can't be made with certainty because the earnings are, at best, expected. Nevertheless, the calculation gives us something to work with and appears to be a good, if not the best, estimate of what will actually occur. We can also use mathematical expectation to compare the expected outcomes of two or more actions. For example if investing a given amount of money in stocks gives an expected profit of $10,000 after one year and investing the same amount in bonds gives an expected profit of $20,000 after one year, a rational investor would invest in bonds.

This chapter includes examples of how mathematical expectation can be used to make such decisions. The diversity of the applications is most unexpected, ranging from gambling to theology. The last two sections involve highly counterintuitive outcomes, serving as a preview of the chapters to follow.

What Happens in Vegas

The 1990s attempt by Las Vegas to promote itself as a family-friendly tourist destination has long since come to an end. In 2002, to resurrect the image of Las Vegas as Sin City, the Las Vegas Convention and Visitors Authority began promoting the slogan, "What happens in Vegas stays in Vegas." We know what this means. But it also applies to money wagered by tourists in Las Vegas casinos. A significant proportion of bets placed are bets lost, then becoming gambling revenue for the casinos. September of 2010 Las Vegas Strip gambling revenues exceeded $500 million (approximately $700,000 per hour). Casinos expect to win and they certainly will in the long run. It follows that gamblers should expect to lose, and lose they do. Mathematical expectations can be calculated for casino games, and in every case the player's expectation is negative, indicating an expected loss in the long run. Do betting strategies exist to produce a positive expectation? If not, why would a rational individual gamble at a casino?

One of the easiest games to analyze is American roulette. The wheel has 38 numbered sectors, or partitions, numbered 1 to 36 (half colored red and half colored black) with the two remaining sectors numbered 0 and 00 (both colored green). The wheel is spun and players wager that a ball will fall into one of the partitions. If the ball lands on the selected number (or one of a set of selected numbers) the player wins; otherwise the player loses. For example, say the player places a $1 bet on number 7. If the ball lands on this number, the player wins $35. If not, the bet is lost. The probability of winning this bet is 1/38 and the player's expectation for a single bet of this type is

$$E = \frac{1}{38}(\$35) + \frac{37}{38}(-\$1) = -\$\frac{1}{19} \approx -5¢.$$

By betting on any single number, as above, the player loses, on average, approximately 5¢ on each $1 wagered, or 5 percent of the bet. After one hundred such bets, the player is expected to have lost approximately $5. Equivalently, the casino wins 5 percent of the money wagered by players. This house advantage holds for most other types of roulette bets. Instead of betting on a single number, the player can bet on combinations of numbers such as odd or even, red or black, etc. If the player places a $1 bet on red and the ball lands on any red number, the player wins $1. Otherwise the player loses the bet. For this bet the player's expectation is

$$E = \frac{18}{38}(\$1) + \frac{20}{38}(-\$1) = -\$\frac{1}{19} \approx -5¢,$$

the same expectation as betting on a single number.

The European roulette wheel has 37 sectors numbered 1 to 36 with the remaining sector numbered 0. Using this layout, the house advantage drops to just under 3 percent.

In the short run, a player can win. But Bernoulli's law of large numbers dictates that, in the long run, the player will lose and the casino will profit. And this is the case for all casino games; otherwise, the casino would lose money and go out of business. For Keno the house advantage can be as large as 27 percent, and for slot machines it varies between 4 percent and 8 percent. For blackjack (21) the house advantage varies as the player's skill is a factor. Using card-counting techniques and playing with a single deck, a player can achieve a 1 percent advantage and theoretically win in the long run. To minimize this, casinos play the game with multiple decks and may ask a player to leave the table if card counting is suspected. So, the only foolproof scheme a player can use to avoid a long-run loss is to simply not play.

The *Martingale* betting system, a scheme that, in principle, favors the player, seems too good to be true. And if it seems too good to be true, ... The idea is to place a $1 bet on any even-money bet, such as betting on red in American roulette. If the best is won, the player can once again bet $1 on red. But if the bet is lost, then the player doubles the bet, wagering $2 on red. If this bet is lost, then the player doubles the wager and bets $4 on red. The player continues to double the wager after each loss, until the ball lands on red, at which point the player wins a large sum of money, making up for the previous string of losses. Whenever a win occurs, the player's account will be $1 more than after their last win.

For example, the player begins by placing a $1 bet on red and loses. The player doubles the bet after each successive loss and eventually wins on the fifth bet of the sequence. The outcomes loss, loss, loss, loss, and win give the player

$$(-\$1) + (-\$2) + (-\$4) + (-\$8) + \$16 = \$1.$$

Since the eventual win is inevitable, the player appears guaranteed to win in the long run. In actuality, the opposite is true. There are two foils causing Martingale schemes to fail.

A player would need a large bankroll to sustain a long sequence of losses. If, for example, the player were to lose ten times in a row, the eleventh bet would require a wager of $1,024. If the player were unable or unwilling to wager this amount, the game would end and the player would have lost $1 + $2 + $4 + ... + $512 = $1,023. Even if the player did have an unlimited bankroll, the casino itself places an upper limit on bets. The maximum allowable bet is typically 100 to 200 times the minimum bet. So, if the player begins by placing the minimum allowable bet of $1 and suffers a sequence of ten losses, the required wager of $1,024 on the eleventh bet might not be allowed and the player would suffer a significant loss.

So, in the short run, the player could win small amounts of money by Martingaling. In the long run, a significant run of losses will occur, bankrupting the player.

The short-run wins are minimal but a similar scheme can be used to increase the player's expectation to $1 per play or, in this case, $1 per spin of the wheel. As before, play begins by betting $1 on an even-money bet, say red. If the player wins, then the $1 wager can be repeated. But if a sequence of $n - 1$ losses occurs, then the player is to wager $2^n - 1$ dollars on the nth wager. When the inevitable win occurs, the gain will equate to winning $1 on each of the n bets. For example, the outcomes loss, loss, loss, loss, and win give the player

$$(-\$1) + (-\$3) + (-\$7) + (-\$15) + \$31 = \$5,$$

giving the player an expected profit of $1 per play. But, bankroll and house limits apply as before and the system will fail in the long run.

Gambling authority John Scarne gives this amusing account of a "successful" roulette system that worked once and will certainly never succeed again [Scarne 74, pp. 414–415]:

> A few years back in a Houston, Texas, casino an elderly, distinguished-looking gentleman slightly in his cups wavered back and forth behind a group of women players at the roulette table. Nobody paid any attention to him until he began complaining about how unlucky he was.
> "What do you mean, unlucky?" the croupier asked.
> "Number 32 just won, didn't it?" the grumbler said.
> "Yes, but you didn't have a bet down. What's unlucky about that?"
> "Oh, yes, I did," the drunk groaned. "I made a $10 mind bet on 26 and lost!" Then he handed the croupier a $10 bill. "I always pay my losses—even on mind bets."
> The croupier tried to return the money, but the old gentleman stubbornly refused to take it. Since this argument was creating a commotion and interrupting the game's action, the croupier finally shrugged, smiled wryly and shoved the bill into the money box.
> The drunk, apparently satisfied, disappeared in the direction of the bar, but he was back again before long. He walked up to the table just as the croupier spun the ball. He wobbled unsteadily and watched until the ball dropped; then he came to life, shouting excitedly, "That's me! I bet ten bucks on number 20 and I won!"
> The croupier tried to continue the play, but the drunk, who suddenly seemed much more sober, interrupted loudly, demanding to be paid the $350 he had won on his mind bet. He kept this up until the casino manager was called. After hearing what had happened, he ruled that since the croupier had accepted a $10 losing mental bet, he must pay off on the winning mind bet. You can be quite sure that this was the last mental bet which that croupier or any other in that casino ever accepted.

If the player's only objective is to win, then no rational player would play casino games where the player's expectation is negative. If the utility or intrinsic entertainment value exceeds the expected loss, then gambling is rationally justified. Otherwise, why do it? It's sad to see players betting their entire paychecks on worthless schemes that they believe will pull them out of poverty. Along these lines, state-run lotteries have been called a "tax on ignorance" by those opposed to using public lotteries as a means of generating state revenue. The description is harsh, but it only targets those who gamble large sums of money and can't afford to do so, hoping to win big. "Someone will win. Why not me?" Someone will win, of course, but the probability of any one individual winning is extremely small. For a typical state-run lottery in the US, the player's expectation for a $1 bet is −50¢, ten times worse than it is for American roulette. Only 50 percent of money is returned to the player in the form of winnings. Approximately 30 percent goes to the state and 20 percent is for operating costs.

Is Insurance a Good Bet?

Purchasing an insurance policy is mathematically equivalent to placing a bet at a roulette table. The amount wagered is the annual premium. The player, or in this case the policy holder, is betting on the possibility of experiencing some sort of misfortune during the term of the policy. The policy holder wins the bet if, in fact, such misfortune were to occur. Otherwise, the bet is lost and the player is free to play again by purchasing the policy for an additional year. Determining the expected value of the policy for the policy holder is mathematically identical to the roulette calculation in the previous section.

For example, suppose a 20-year-old female purchases a $250,000 one-year term life insurance policy for $200. National vital statistics indicate she will survive the year with probability .999546. That is, if she dies, the insurance company pays a $250,000 death benefit to her heirs. (To be precise, the actual payoff is $250,000 − $200 = $249,800 since the premium is not returned.) If she survives, the insurance company keeps the $200 premium and she gets nothing. From her perspective, the expected value of the policy is

$$E = .000454(\$250,000 - \$200) + .999546(-\$200) = -\$86.50.$$

Equivalently, the insurance company expects to earn, on average, $86.50 per policy sold. The "game" favors the insurance company in the same way that roulette favors the casino. Ignoring additional considerations, it's a bad bet and, in the long run, she would be better off not purchasing the policy and investing $200 per year elsewhere.

Much like all casino games show negative player expectations, the same holds for insurance. If this were not so, the insurance companies would lose money and financially fail. They provide a service and to survive they must make a profit. One may ask why any rational individual would purchase an insurance policy if, in the long run, policy holders pay out more than they receive. At least with casino gambling there is the intrinsic value of entertainment. But what fun is there in paying annual premiums and hardly ever getting anything back? And winning a life insurance bet is nothing to celebrate either. It's literally a bet policy holders are "dying to win."

There are three reasons why, under certain conditions, it makes sense to purchase insurance. The primary reason is to protect the policy holder against a catastrophic financial loss that would either be unsustainable or cause the uninsured extreme hardship. Life, medical, and automobile liability policies are desirable because a death, hospitalization, or serious traffic accident could be both physically as well as financially catastrophic. On the other hand, an individual's financial status may be such that a given loss would have little relative financial impact. For example, there would be no need for a billionaire to purchase vision care insurance (optometric services including eye exams, frames, lenses, etc.) since care costs would be insignificant when compared to the individual's financial resources.

The second compelling reason to purchase insurance is that it may be required by law to do so. The vast majority of states in the US have some form of compulsory automobile liability insurance where failure to have such insurance may result in the suspension of a driver's license. The purpose of mandatory auto insurance is to protect the public (both driver and victim(s)) from the high medical and property damage costs associated with automobile accidents.

On March 23, 2010, US President Barack Obama signed into law a health care bill that, by the year 2014, requires most Americans to have health care insurance. Employers would be required to provide coverage to workers. Those refusing to comply would be fined. Low-income Americans and companies with fewer than 50 employees would be exempt. There is some controversy, as some argue the mandate is in violation of one's personal freedom to choose whether or not to purchase insurance. Some conservatives feel the federal government overstepped its authority.

The least justifiable, yet often compelling, reason to purchase insurance is the peace-of-mind factor. If the possibility of a financial loss causes an individual to lose sleep and become physically ill, then it makes sense to purchase insurance and eliminate the risk of loss. This may be an irrational decision based on unfounded fears; yet, in some cases, the peace of mind gained by purchasing the insurance may exceed the cost of the premium. And what is the fair price of peace of mind? This depends on the individual's state of mind, which may be entertaining irrational concerns.

Insurance companies promote the peace-of-mind factor when marketing policies. The cupped "good hands" iconic logo of Allstate Insurance evokes images of a warm, caring company, providing a financial safety net in the event of a loss. The family provider is subliminally encouraged to do the right thing by protecting loved ones against all possible loss. "Now you can rest comfortably knowing ...." Similar feelings of well being and security are suggested by the umbrella logo of Travelers Insurance and the Rock of Gibraltar used by Prudential Insurance.

Mathematical expectation gives prospective policy holders a means of comparing policies in order to select a best buy. Consider the earlier example of the

20-year-old female who purchases a $250,000 one-year term life insurance policy for $200. Her expectation is −$86.50 based on her probability of surviving the year. She compares this policy to one offered by another insurance company where she can purchase a $400,000 one-year term life insurance policy for $250. A calculation similar to the previous one shows an expectation of −$68.40, making it a slightly better buy than the original policy. Of course, she must consider whether or not she can afford the higher premium and determine if she really needs the extra protection. Those considerations aside, the second policy is preferable to the first.

If there are only three justifications for purchasing an insurance policy, then it would appear that many types of policies are purchased unnecessarily by policy holders believing the purchase was mathematically justified. The best example may be the extended warranty option, offered to customers when they purchase a home, car, or appliance. The policies are associated with high profit margins (between 40 and 80 percent) and are aggressively marketed to customers at a time when they are most vulnerable. Some customers reason, "Why not spend an additional $50 to double the term of my warranty for this $500 appliance?" Relative to $500, the $50 premium seems small. Yet, the expectation on such a bet is negative, and a required repair of the appliance may not be as catastrophic as it would appear at the time of purchase. In addition, there may be a high deductible and the warranty may not cover all the consumer believes it does. And then there's the problem of double coverage. The extended warranty offered by the store selling the item may run concurrently with the manufacturer's warranty, in which case one of the two policies is worthless. Buyer beware!

Consumers face a double-coverage dilemma when renting a car and the agent offers the option of purchasing additional insurance. The renter's own automobile insurance may already provide sufficient coverage as may the renter's credit card used for the rental. Many renters purchase the additional insurance unnecessarily, while standing at the airport rental counter in a hurry to leave the airport and get on with their plans. If they don't know what kind of coverage they already have, there's a good chance they will play it safe and buy the additional insurance in order to be certain they are covered. The insurance they purchase may be effectively worthless if they're already covered and the cost is high, often as much as the rental rate for the car alone.

Airline Overbooking

Year after year there are more passengers being denied boarding on a flight for which they have reservations. *Bumping* occurs as a consequence of overbooking, the practice of a carrier booking more seats than are available on a given flight. It's done to account for predicted no-shows, passengers with reservations that never show up at the airport and have not canceled their reservation. The intent of overbooking is to keep planes flying full, not to waste empty seats. Legal? Yes.

Bad for passengers? Not necessarily. Bumped passengers are financially compensated in the form of seat upgrades and vouchers on future flights. The compensation is significant enough that some passengers volunteer to give up their seats in exchange for the compensation. Other passengers will go so far as to intentionally check in late, hoping to be bumped.

Carriers need full flights to maximize profits. Money is lost on empty seats and these costs are passed on to the flying public. So, the airlines do what they must do and not everyone is terribly disturbed by this fact of life.

Mathematical models of overbooking have been published with the intent of determining the optimal number of passengers that should be booked for a given flight to maximize the carrier's profit. The no-show factor is taken into account, and the model almost always recommends a degree of overbooking. It's not an exact science and the amount of overbooking varies by carrier, route, and time of travel. At the time of this writing, Ryanair, a low-cost Irish airline, claims to be the only airline in Europe that does not overbook its flights, eliminating the possibility of passenger bumping. Presumably this is done to promote good customer relations, which, in turn, is good for business. This is the same airline that is promoting the installation of coin-operated pay toilets on their fleet and the removal of ten seats on each aircraft to allow for 15 standing passengers. So much for customer relations!

The model to follow makes the following simplifying assumptions:

1. Assume the carrier's fleet consists of aircraft, each with a seating capacity of C passengers, not including the pilot and other flight personnel.
2. The price of a ticket is the same for each passenger.
3. The carrier maximizes revenue on each flight by flying full with C passengers. We assume operational and fuel cost are the same for any given flight, and do not depend on the number of passengers that board. This is, of course, not entirely true as each additional passenger increases fuel and other operational costs. But, we take these costs to be relatively small when compared to the price of a ticket.
4. Passengers may reserve a seat, in advance, with no required deposit. So, if they fail to show up at the airport, they suffer no financial penalty.
5. The probability of a randomly selected passenger with a reservation showing up is p, in which case the probability of this passenger being a no-show is $q = 1 - p$.
6. The probability of exactly k passengers arriving at the airport for a flight with n existing reservations is given by the binomial probability,

$$\frac{n!}{k!(n-k)!}p^k q^{n-k}.$$

Number of Reservations n	Expected Revenue (Rounded)
0	$0
1	$70
2	$140
3	$210
4	$280
5	$350
6	$420
7	$474
8	**$497**
9	$490
10	$458

Table 3.1. Expected revenue as a function of the number of reservations.

7. Any passenger denied boarding caused by overbooking will be compensated with a cash payment or voucher that can be used on a future flight. We assume this compensation is the same for every such bumped passenger.

As an example, consider a carrier offering 30-minute helicopter sightseeing flights over the Las Vegas Strip. Each helicopter of the fleet can hold up to six passengers ($C = 6$), as well as the pilot. The ticket price per passenger is set at $100. The probability that any given passenger with a reservation shows up is taken to be $p = .7$ with a no-show rate of $q = .3$ or 30 percent. If the carrier books to capacity but not beyond ($n = C = 6$), then the expected number of arrivals for the flight will be $.7(6) = 4.2$, corresponding to one or two expected empty seats. So, the carrier decides to overbook ($n > C$) offering any bumped passenger a free flight voucher or the equivalent cash compensation of $100. How many passengers should be booked to maximize the expected revenue?

Clearly n should be at least equal to, if not greater than, C. If a flight is booked to capacity by taking $n = C = 6$ reservations, the expected revenue is given by summing the products of binomial probabilities and their associated revenues:

$$E = \sum_{k=0}^{6} \frac{6!}{k!(6-k)!}(.7)^k (.3)^{6-k} \cdot \$100k = \$420.$$

If the carrier overbooks by one passenger ($n = 7$), then they would suffer a \$100 penalty only if all seven of the passengers show up. Overbooking by one passenger gives an expected revenue of

$$E = \sum_{k=0}^{6} \frac{7!}{k!(7-k)!}(.7)^k (.3)^{7-k} \cdot \$100k + \frac{7!}{7!(7-7)!}(.7)^7 (.3)^0 (\$600 - \$100) \approx \$474.$$

Table 3.1 gives the expected revenue for $0 \le n \le 10$, under the above assumptions.

Based on calculations summarized in Table 3.1, the carrier should attempt to book eight reservations on each of the six passenger helicopter flights. In doing so, it runs a risk of overbooking passengers to whom must be paid denied-boarding compensations. It's a risk worth taking.

Interestingly, the optimal value of $n = 8$ is independent of the ticket price. If the price of a ticket is doubled to \$200 (in which case a denied-boarding voucher or cash equivalent also doubles), the maximum expected revenue doubles, but still corresponds to $n = 8$ reservations. This is easily seen from the previous expected revenue calculations.

But what if the denied-boarding compensation were to change with the ticket price of \$100 remaining constant? If the compensation increases, a lesser degree of overbooking would be warranted. If, for whatever the reason, the denied-boarding compensation goes up to \$750, then an optimal expected revenue of \$460 would be achieved by booking to capacity ($n = 6$) with no overbooking. From the carrier's point of view, the steep denied-boarding compensation makes it unfeasible to overbook and risk the severe penalty.

On the other hand, if the denied-boarding compensation is reduced and the ticket price remains at \$100 per passenger, the carrier's overbooking penalty decreases and a higher degree of overbooking would be suggested. In the extreme case where denied-boarding compensation is eliminated, the carrier should book as many seats as possible on any given flight as it would have nothing to lose no matter how many passengers show up. Under these conditions, almost every flight would depart full with the expected revenue per flight nearing \$600. Carriers can't afford to do this as prospective passengers would be wary of being bumped without compensation and the booking demand may fall off. This, along with the bad public relations generated, might force the carrier to reduce the number of flights and suffer a decrease in total profit. And, it may be that this form of unlimited overbooking is illegal.

Another option to consider would be for the carrier to offer a given number of seats with a prepay "use it or lose it" reservation. Those passengers arriving at the airport with a prepaid ticket would be guaranteed a seat no matter how many other passengers show up. Say, for the six-passenger aircraft under consideration, two seats are sold on this basis. Then to maximize the revenue for the flight, one need only maximize the expected revenue for the four remaining seats being reserved with no prepayment and a potential denied-boarding compensation. The analysis would be similar to what has been previously presented.

Additional considerations would make this model more accurate. But, the mathematics would be too technical to present here. There's also the psychology of the ticket buyer to consider, where unexpected revenue expectations may appear in the form of unintended consequences.

As an example, assume the carrier books seats with no prepayment and begins to suffer losses as a result of the no-shows. It's proposed that, at the time of booking, a customer's credit card number be recorded with the understanding that there will be a small nominal charge if they fail to show. The charge would be minimal, say $5, and its only purpose would be to discourage no-shows. Would this reduce the number of no-shows and increase revenue for the carrier?

It might. But then again, things could backfire. Consider the published study of ten day-care centers in Haifa, Israel, where the operators were having problems with parents not showing up on time to pick up their children [Gneezy and Rustichini 00]. To minimize the problem, it was decided to charge parents a small penalty of ten shekels ($\approx$ $3) per child if a parent were to arrive more than ten minutes late. Conventional wisdom suggests the late arrivals would decrease as a consequence of the penalty. In actuality, the opposite happened with more parents arriving late for their children. The authors of the study attribute the increase in tardiness to the elimination of the moral incentive to arrive on time. Some parents may have felt that the ten shekel fee legitimized the tardiness and were willing to pay for the privilege of arriving late, without guilt.

Returning to our Las Vegas scenic flight example, would the $5 charge discourage no-shows? Human nature is difficult to predict, and only through trial and error can such a question be answered.

The above analysis assumes a fixed value payment or voucher for anyone denied boarding caused by overbooking. A different and more complex analysis is required if the compensation is negotiable. In late 2010 Delta Airlines introduced a new high-tech scheme for compensating bumped passengers, designed to eliminate confusion at the gate for oversold flights and to reduce the costs of compensation. During the check-in process at airport kiosks, passengers are asked to enter the minimum dollar amount they would accept to be bumped. If the flight is overbooked, Delta can then accept the lowest bids, which could presumably be lower than the standard voucher, currently ranging from $200 to $400. The plan is promoted by Delta as a time saver. The airline acknowledges it will also save money by offering compensation to the lowest bidders.

Composite Sampling

A medical lab tests the blood of five individuals for a blood-borne pathogen that effects 1 percent of the population. The five individual blood samples are sent to the lab. If each sample is individually tested, five tests are required. Another

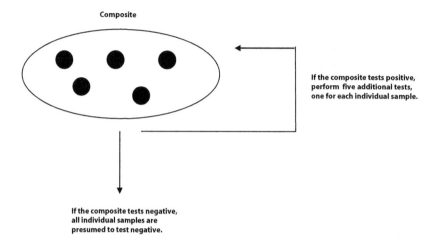

Figure 3.1. Composite blood testing.

option would be to pool the five blood samples forming one composite. If the composite tests negative then all five individuals are assumed to be pathogen free and no additional testing is required. On the other hand, if the pathogen is detected and the composite tests positive, then each individual sample would have to be tested to determine exactly which of the samples are effected. In such a case a total of six tests would be performed, the original composite test plus five individual tests. (See Figure 3.1.)

If the lab uses the second option of pooling the samples, what is the expected number of tests that would have to be performed? For the composite to test negative, the pathogen would be absent in each of the individual samples. This occurs with probability $(.99)^5 \approx .95$; the composite would test positive with probability $p \approx .05$. The expected number of tests required is therefore

$$E \approx .95(1) + .05(6) = 1.25.$$

Using composite sampling to test the blood dramatically reduces the expected number of tests from five (if the samples are tested individually) to just over one, thereby reducing cost and the time required to report the test results. Rather than requiring one test per individual, the composite sampling method requires, on average, approximately 0.25 tests per individual, a dramatic improvement in efficiency.

If a large number of individuals require testing, we should be able to find the optimal number of samples to pool in order to minimize the expected number of tests per individual. As shown in Figure 3.2, using 11 samples in each composite will minimize the expected number of required tests per individual to approxi-

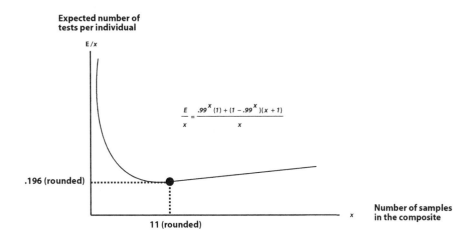

Figure 3.2. Minimizing the expected number of tests per individual.

mately 0.196, an eighty percent reduction when compared to testing each sample individually.

A classic application of composite sampling occurred during World War II when US servicemen were tested for syphilis by testing their blood for a specific antigen of the syphilis-causing bacterium. Today the technique is also used for testing soil, water, and air for environmental hazards.

There are caveats. For pooling to be feasible, the condition being detected must have a low prevalence. In the previous example, we assumed a prevalence of 1 percent. It would make sense to pool samples in this situation as there is a high probability that no more than a single test (with a negative result) would be required. Compare this to testing for a condition having a prevalence of 50 percent. When two samples are pooled, the test result would be positive at least 75 percent of the time, requiring the testing of individual samples. The most efficient testing procedure would be to forego the testing of a pooled sample and go straight to the testing of individual samples. And how low is low? Pooling is feasible if roughly less than 30 percent of the samples to be tested have the condition in question. (The exact cutoff is given by the odd expression $1 - 1/\sqrt[e]{e} \approx .31$, where e is the natural logarithm base. For this prevalence, exactly e (unrounded) samples are required for the composite. The derivation is given in this chapter's appendix.)

Composite sampling may be problematic if mixing individual samples changes the chemical nature of a sample. This can occur if sample constituents interact. And then there's the problem of dilution. Detection of a pathogen, antigen, or some contaminant may require that its density exceed a certain threshold. If mixing samples significantly lowers this density for otherwise positive testing samples, composite sampling should not be used.

Pascal's Wager

If one's belief in a deity is based on faith alone, then a rational justification for the belief may not exist. In nine words, "God said it. I believe it. That settles it." And indeed it does for the believer. Others require a reason to believe.

Blaise Pascal offers such a reason, in the form of an argument as to why one should wager in favor of God's existence. *Pascal's wager*, one of the most famous arguments in theology, compares the consequence of believing to that of not believing. Pascal begins by making the assumption that if there is a God, we are incapable of knowing so with certainty. His argument begins [Pascal 05, pp. 40–41],

> Let us examine the point and say, "God is, or He is not." But to which side shall we incline? Reason can decide nothing here. There is an infinite chaos which separates us. A game is being played at the extremity of this infinite distance where heads or tails will turn up. According to reason, you can do neither the one thing nor the other; according to reason you can defend neither of the propositions.
>
> Do not reprove for error those who have made a choice; for you know nothing about it. "No, but I blame them for not having made this choice, but a choice; for again both he who chooses heads and he who chooses tails are equally at fault, they are both in the wrong. The true course is not to wager at all."
>
> —Yes; but you must wager. It is not optional. You are embarked.

Pascal's argument is mathematical. The consequences (payoffs) associated with believing or not believing are given in Matrix 3.1 as a payoff matrix.

If God exists, we are far better off wagering for God's existence. If God doesn't exist, then the outcome is insignificant one way or the other. One has much to gain and nothing to lose by believing in God.

Quantification of the above payoffs is somewhat arbitrary. Here we assign ∞ to eternal bliss and $-\infty$ to damnation. The symbols are used to represent ex-

	God exists	God doesn't exist
Believe in God	Eternal bliss	Minimal effect
Don't believe in God	Damnation	Minimal effect

Matrix 3.1. Pascal's wager consequences.

	God exists	God doesn't exist
Believe in God	∞	c_1
Don't believe in God	$-\infty$	c_2

Matrix 3.2. Pascal's wager payoff matrix.

tremely large, if not infinite utilities. We choose small finite numbers c_1 and c_2 to represent the two minimal consequences associated with God's nonexistence. Matrix 3.2 is the numerical payoff version of Matrix 3.1.

If we let p_1 denote the probability that God actually exists and p_2 the probability God does not exist, then one's expectation in believing in God is given by

$$E_{\text{believer}} = p_1 \cdot \infty + p_2 \cdot c_1 = \infty,$$

taken to be a large, if not infinitely large positive expectation. The expectation for a nonbeliever is given by

$$E_{\text{nonbeliever}} = p_1 \left(-\infty\right) + p_2 \cdot c_1 = -\infty,$$

taken to be a large, if not infinitely large negative expectation. So $E_{\text{believer}} > E_{\text{nonbeliever}}$ regardless of how likely it is that God actually exists. If rational behavior is taken to mean acting in such a way as to maximize expectation, then a rational individual should, according to Pascal's argument, wager that God exists and lead a pious life. Science philosopher Ian Hacking calls the wager the first well-understood contribution to decision theory [Hacking 06, p. 62].

As expected there are many objections to the argument. For starters, the entire argument could be dismissed as a trivialization of religion, reducing it to nothing more than a game where one plays to win. The argument fails as well for strict atheists. They would assign a probability of zero to God's existence ($p_1 = 0$) in which case $p_2 = 1$. Then $E_{\text{believer}} = c_1$ and $E_{\text{nonbeliever}} = c_2$. Since we are not assuming $c_1 > c_2$, the argument fails. Others take issue with the concept of infinite expectation. Even if the payoff were quantified as being infinite, our sensations are finite and the intrinsic worth of the payoff must be finite as well.

Famed evolutionary biologist and atheist Richard Dawkins strongly objects to Pascal's wager argument [Dawkins 06, pp. 130–132]:

> Believing is not something you can decide to do as a matter of policy. ... Pascal's wager could only ever be an argument for *feigning* belief in God. And the God

that you claim to believe in had better not be of the omniscient kind or he'd see
through the deception. ... Would you *bet* on God's valuing dishonestly faked belief (or even honest belief) over honest skepticism?

The logic of Pascal's wager appears in H. G. Well's *Apropos of Dolores*, where the
narrator, Stephen Wilbeck, defends optimism [Wells 08, p. 25]:

> While there is a chance of the world getting through its troubles I hold that a
> reasonable man has to behave as though he was sure of it. If at the end your
> cheerfulness is not justified, at any rate you will have been cheerful.

Game Theory

Game theory received no significant world attention until the publication in
1944 of *Theory of Games and Economic Behavior* by American mathematician
John von Neumann (1903–1957) and Austrian economist Oskar Morgenstern
(1902–1977). It was sixteen years prior that von Neumann proved the minimax
theorem, the fundamental theorem of game theory. Years later, the mathematician and the economist produced the groundbreaking collaboration, regarded
as one of the major scientific achievements of the twentieth century.

A *game* is an interaction of two or more individuals (players), each choosing options in such a way as to win as much as possible. If there are only two
competing forces, the games are called *two-person* games. Examples include
parlor games (chess, checkers), a variety of sporting events (football, basketball),
labor-management negotiations, and military conflicts between two nations. If
one player's gain (as denoted by a positive number) equates to the other player's
loss (as denoted by a negative number), then after playing the game the sum of
the players' gains is zero and the game is called a *zero-sum* game. Non-zero-sum
games are discussed in Chapter 9. This section will be primarily limited to *two-
person, zero-sum* games.

A player may choose a pregame strategy, or prescription, for how to play
in any given situation that may arise. Any strategy is legal, as long as it does not
violate the rules of the game. A rational player will attempt to use an *optimal
strategy*, one that maximizes the expected win (gain) per game, minimizing the
expected loss per game. A two-person, zero-sum game is called *solvable* if optimal strategies can be calculated for both players. The *solution* is taken to be the
players' optimal strategies along with the expected gain for a player (loss for the
other) when optimal strategies are adopted.

Zero-sum games, where each player has a finite number of available moves
at each play, can be modeled in either *normal* (matrix) form or *extensive* (tree)
form. Normal form is more compact. Extensive form allows for the consideration
of additional game characteristics.

Beginning with a simple example, we assume Ron and Carla each hold a nickel (5¢) and a quarter (25¢) in their hands. They each simultaneously place one of

Carla

	5¢	25¢
Ron 5¢	−5¢	25¢
25¢	5¢	25¢

Matrix 3.3. A nickel-quarter coin game in normal form.

their coins on the table. If the sum exceeds 20¢, Ron wins Carla's coin; otherwise, Carla wins Ron's coin. In normal form, the game is presented as a matrix showing all possible moves for Ron and Carla with the associated payoff for Ron. A positive entry is a win for Ron and a negative entry is a loss. Since this is a non-zero-sum game, there is no need to show Carla's payoffs. They are opposite in sign to Ron's payoffs. Matrix 3.3 shows the payoff matrix for this game.

In terms of the matrix, think of Ron as the row player and Carla as the column player. Ron chooses a row as Carla simultaneously chooses a column. Ron's payoff is given in the corresponding cell. So, if Ron shows the nickel and Carla shows a quarter, Ron wins 25¢ from Carla. If they each show a nickel Ron loses his nickel to Carla.

The extensive form treats the moves as being sequential, whether or not they are. The game depicted in Matrix 3.3 can be made sequential by allowing Ron to choose his coin first and then allowing Carla to choose, without knowing

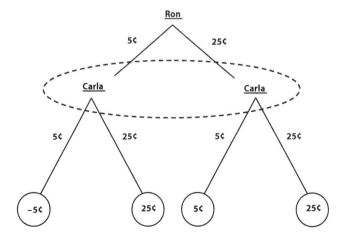

Figure 3.3. A nickel-quarter coin game in extensive form.

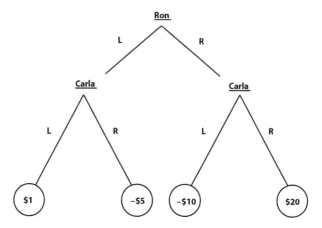

Figure 3.4. A left-right game in extensive form.

Ron's choice. Figure 3.3 shows the extensive form of this game, with Ron's payoffs shown at the bottom.

The dashed oval represents Carla's lack of information as to her position in the tree.

Surprisingly, sequential games in extensive form can be converted to normal form, even when the players are given full information about all previous moves. Games of perfect information include chess, checkers, and tick-tack-toe (naughts and crosses). As an example, Ron and Carla play a game where Ron moves first, choosing left (L) or right (R). After seeing Ron's move, Carla chooses between the same two options. If Ron goes left and Carla does the same, Ron wins $1. If Ron goes left and Carla goes right, Carla wins $5. If Ron goes right and Carla goes left, Carla wins $10. Finally, if Ron goes right and Carla does the same, Ron wins $20. The extensive form of this game and the payoffs to Ron are given in Figure 3.4. Note there is no dashed oval as both players have perfect information about all previous moves.

To convert this game to normal form, we can think of the players as simultaneously choosing strategies. Ron chooses a strategy of left or right. Carla simultaneously chooses a strategy from one of four options:

1. Go left if Ron goes left and go left if Ron goes right (L if L and L if R).
2. Go left if Ron goes left and go right if Ron goes right (L if L and R if R).
3. Go right if Ron goes left and go left if Ron goes right (R if L and L if R).
4. Go right if Ron goes left and go right if Ron goes right (R if L and R if R).

The normal form payoff matrix is shown in Matrix 3.4.

As the normal form is more compact, we will use it to model the games in this chapter and the non-zero-sum games in Chapter 9.

Carla

	L if L and L if R	L if L and R if R	R if L and L if R	R if L and R if R
Ron L	$1	$1	−$5	−$5
Ron R	−$10	$20	−$10	$20

Matrix 3.4. A left-right game in normal form.

A strategy for the row player of a game in normal form is the set of probabilities the row player will use to play the various rows. A strategy for the column player is defined similarly. For the nickel-quarter coin game shown in Matrix 3.3, Ron may randomly select his coin, so that the probability of choosing either coin is 1/2. Carla may choose the strategy of selecting the nickel with probability 1/5 and the quarter with probability 4/5. The outcome of any single play is uncertain as the strategies chosen are *mixed*, allowing for the possibility of either coin being chosen. If Ron and Carla employ these strategies, then the expected payoff is

$$E = \frac{1}{2}\cdot\frac{1}{5}(-5\text{¢}) + \frac{1}{2}\cdot\frac{4}{5}(25\text{¢}) + \frac{1}{2}\cdot\frac{1}{5}(5\text{¢}) + \frac{1}{2}\cdot\frac{4}{5}(25\text{¢}) = 20\text{¢}.$$

On average, Ron will win 20¢ per play of the game and Carla will, on average, lose 20¢ per play. Ron could employ a *pure* strategy of playing the nickel with probability 1 (always) and playing the quarter with probability 0 (never). If Carla adopts the same pure strategy of always playing the nickel, then the expected (guaranteed) payoff is $E = -5$¢. Carla wins 5¢ each time the game is played.

Is the game a fair game? If not, who does it favor? To answer the question we must solve the game by determining the optimal strategies for each player and the expected payoff when the optimal strategies are employed. A game for which optimal strategies are pure is easily solved. The nickel-quarter game is one such example. A quick analysis of the payoff matrix shows that no matter which coin Carla selects, Ron can do no worse by choosing row 2, the quarter. This is so because each payoff in row 2 is greater than or equal to the corresponding payoff in row 1. We say row 2 *dominates* row 1 and call row 1 *recessive*. We can delete row 1 with respect to the analysis of optimal strategies. So, Ron's optimal strategy is pure and he should play the quarter 100 percent of the time. Similarly for Carla, she can delete column 2 as being recessive. This is so since each payoff in column 1 is less than or equal to the corresponding payoff in column 2. We say column 1 dominates column 2 (Carla prefers column 1 to column 2), call column 2 recessive, and delete it. Her optimal strategy is to always play the nickel. Because the optimal strategies are pure, the game is solved and strictly determined, in

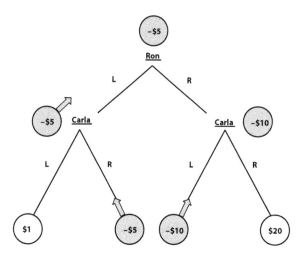

Figure 3.5. Backward induction showing the solvability of the left-right game.

the sense that the outcome is predetermined if both players play optimally. Ron will win 5¢ at each play of the game and Carla will lose this amount. The *value* of a game is its expected payoff when optimal strategies are employed. Letting *V* denote this amount we have *V* = 5¢. Any game such that *V* > 0 favors the row player. Games such that *V* < 0 favor the column player. A game is defined as *fair* when *V* = 0.

A quick way to spot a normal form game that is strictly determined (having pure optimal strategies) is to look for a *saddle point*,[1] a payoff that is simultaneously a minimum in its row and a maximum in its column. In Matrix 3.3, the row 2, column 1 payoff of 5¢ is a saddle point. Games with saddle points are instantly solved. The row player should play the corresponding row, the column player should play the corresponding column, and the value of the game is the saddle point value. The left-right sequential game is now easily solved by looking at its normal form payoff matrix. Ron should play row 1 by always going left and Carla should adopt the pure strategy corresponding to column 3. The game is strictly determined and Carla will win $5 each time the game is played.

Of interest is the fact that any sequential game where both players have perfect recall of previous moves is strictly determined. That is, games like tick-tack-toe, checkers, and chess are all solvable and have saddle points in their payoff matrices. To see this we need not construct the astronomically large normal form payoff matrices. We need only make a simple backward induction argument on the extensive form of such games.

---

[1] Consider the shape of a saddle, which curves upwards along the spine of the horse and curves downwards around its midsection. The middle point, where the rider sits, is the highest point of the midsection curve but the lowest point of the spine curve. The term *saddle point* is used in mathematics for such a point on a surface as well as for such a point of data in a matrix.

We take the left-right sequential game as an example and work from the bottom up in Figure 3.5, starting with the possible payoffs.

Carla, being the last to play, will choose right if Ron has chosen left and choose left if Ron has chosen right. Remember, Carla prefers negative payoffs to positive ones as they represent losses for Ron and wins for her. Since Carla will make these choices every time the game is played, promote them up the chart as shown. If Ron goes left, his payoff will ultimately be a loss of $5. If he goes right his payoff will be a loss of $10. Rationally he should always go left, followed by Carla's going right. Optimal strategies are pure, the game is strictly determined, and its value is $V = -\$5$, favoring Carla. There's really no point in rational players playing the game because the outcome will be the same every time they play.

A discussion of the solvability of tick-tack-toe, checkers, and chess is given in this chapter's appendix.

But what about games without saddle points that are not so easily solved? John von Neumann's famous minimax theorem, the fundamental theorem of game theory, asserts all two-person, zero-sum finite games are solvable. Playing conservatively, a player chooses a strategy, perhaps mixed, so that among all possible strategies, the negative effect of the other player's best counterstrategy is minimized. An optimal strategy according to the minimax theorem minimizes the maximum damage the other player can cause. Specifically, the column player chooses a strategy so as to minimize the maximum possible value of E (the expected payoff for the row player) no matter what strategy is used by the row player. This is called the column player's *minimax strategy*. Call the associated payoff $E_{minimax}$. The row player chooses a strategy so as to maximize the minimum possible value of E, no matter what strategy the column player uses. This is called the row player's *maximin strategy*, and we can denote the associated payoff as $E_{maximin}$. The minimax theorem claims that $E_{minimax} = E_{maximin}$ and the value of the game is $V = E_{minimax} = E_{maximin}$. The optimal strategy for the row player is the maximin strategy, and the optimal strategy of the column player is the minimax strategy. For games not having saddle points, optimal strategies are mixed. Algorithms and formulas exist for determining mixed optimal strategies. Sometimes mixed optimal strategies are intuitive.

Most readers are familiar with the game *rock-paper-scissors* or *rochambeau*. Here we need only be concerned with the fact that this is a two-person, zero-sum game with rock beating scissors, scissors beating paper, and paper beating rock (see Matrix 3.5).

Anyone playing this game intuitively discovers that the optimal strategy for both players is to randomly choose moves, each with probability 1/3. This is so as a consequence of the game's symmetry. The value of the game is 0, reflecting its fairness. Symmetry alone would suggest this. If Ron adopts an irrational strategy, say that of playing rock 100 percent of the time, Carla could exploit the irrationality by playing paper more often, giving Ron an expected loss and Carla an expected gain.

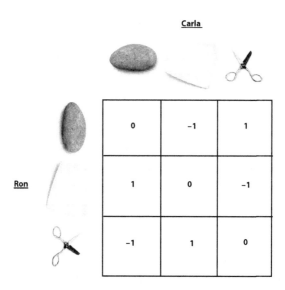

Matrix 3.5. Rock-paper-scissors.

For games having two-by-two payoff matrices without saddle points (see Matrix 3.6), there are simple formulas giving the optimal strategies and the value of such games [Gilligan and Nenno 79, p. 344]. Their derivation is omitted here.

The row player's optimal strategy is to play row 1 with probability $\frac{d-c}{a+d-b-c}$ and row 2 with probability $\frac{a-b}{a+d-b-c}$. The column player's optimal is to play column 1 with probability $\frac{d-b}{a+d-b-c}$ and column 2 with probability $\frac{a-c}{a+d-b-c}$. The value of the game is $\frac{ad-bc}{a+d-b-c}$. So, for the game depicted in Matrix 3.7, Ron should play row 1 with probability 9/10 or 90 percent and row 2 with probability 10 percent. Carla should play column 1 with probability 9/10 or 90 percent and column 2 with probability 10 percent. The value of the game is −1/10 and it favors Carla.

	Carla	
	Column 1	**Column 2**
Row 1	*a*	*b*
Row 2	*c*	*d*

Ron

Matrix 3.6. A general two-by-two game with no saddle point.

Matrix 3.7. A two-by-two game with no saddle point.

Why it is optimal for Ron to play row 1 90 percent of the time when row 1 offers Ron little when compared to row 2? Carla understands the potential loss for herself by playing column 2 and therefore plays column 1 90 percent of the time. Ron understands she will do this and therefore, to minimize his losses, favors row 1 over row 2.

The St. Petersburg Paradox

The St. Petersburg paradox is the best known of all mathematical expectation paradoxes. A brief discussion is included in this chapter because it is an example of a correct, but counterintuitive, calculation; it is not a paradox in the true sense of the word.

A player wages a fixed amount of money and then tosses a fair coin repeatedly until heads first appears. If n tosses are required, the player receives 2^n. So, there is a probability of 1/2 of winning \$2, a probability of $(1/2)(1/2) = 1/4$ of winning \$4, and in general a probability of 2^{-n} of winning 2^n. How many dollars should the player be willing to wager to play the game?

The expected value of the payoff is given by

$$E = \frac{1}{2}(\$2) + \frac{1}{2} \cdot \frac{1}{2}(\$4) + \frac{1}{2} \cdot \frac{1}{2} \cdot \frac{1}{2}(\$8) + \ldots = \$1 + \$1 + \$1 + \ldots = \$\infty.$$

So, a rational player should be willing to wager any finite amount of money and the game still favors the player, who will come out ahead in the long run.

This is not a paradox in the true sense. The conclusion is correct. But are you willing to pay any finite amount of money to play? Would you be willing to wager \$10? Would you wager \$1,000? In theory you should. But few of us would be willing to wager this high amount and an explanation is called for.

The problem was originally proposed by Nicholas Bernoulli and later discussed by his cousin Daniel. There are two popular explanations for the apparent paradox.

The first assumes that arbitrarily large payoffs are impossible since any one individual, casino, bank, or the entire world for this matter has only a finite amount of money. So, if the game were played, there would be a number of tosses after which the payout would remain the same, out of necessity. For the sake of argument, let's say that if 20 or more tosses are required, the payoff must remain constant at 2^{20} = $1,048,576.

Now the expected value of the payoff is given by

$$E = \frac{1}{2}(\$2) + \frac{1}{2} \cdot \frac{1}{2}(\$4) + \frac{1}{2} \cdot \frac{1}{2} \cdot \frac{1}{2}(\$8) + \ldots + \left(\frac{1}{2}\right)^{20}(\$2^{20}) + \left(\frac{1}{2}\right)^{21}(\$2^{20}) + \left(\frac{1}{2}\right)^{22}(\$2^{20}) + \ldots$$

$$= \$20 + \left(\frac{1}{2^{21}} + \frac{1}{2^{22}} + \frac{1}{2^{23}} + \ldots\right)(\$2^{20}) = \$20 + \frac{1}{2^{20}}\left(\frac{1}{2^1} + \frac{1}{2^2} + \frac{1}{2^3} + \ldots\right)(\$2^{20})$$

$$= \$20 + \frac{1}{2^{20}}(\$2^{20}) = \$20 + \$1 = \$21 .$$

Under these conditions, the player should be willing to wager no more than $21 and the paradox disappears.

This explanation lacks something as it skirts the issue by changing the rules of the game. A more satisfying explanation is given by Daniel Bernoulli in terms of the utility of the payoff to the player. In general, the utility of a payoff (the payoff's intrinsic worth to the player) increases with the payoff, but at a decreasing rate. This is the notion of *diminishing marginal utility*. Logarithmic functions increase at a decreasing rate, and for this case we conveniently choose the base 2 logarithm as the player's utility function. For a payoff of size 2^n the utility is $\log_2(2^n) = n$, in which case the player's expected utility is given by $E = \sum_{n=1}^{\infty} 2^{-n} n = 2$. A "proof" that this infinite series converges to 2 is given in this chapter's appendix. The sum is finite and the paradox disappears. The expected utility is low because we're using an oversimplified and somewhat unrealistic utility function. A more realistic function might involve a large constant multiple of this logarithm, but the expected utility would still converge to a finite value, removing the paradox.

Paradox resolved? Not completely. The original problem specifies a payoff of 2^n if heads occurs on the nth toss. If we change the payoff to be 2^{2^n}, then the utility of this payoff is $\log_2(2^{2^n}) = 2^n$ and the expected utility to the player is again infinite.

Stein's Paradox

Averaging is a fundamental process in statistics, second only to counting. Simple averaging sums the data and divides by the number of values. The calculated value is the average, or *mean*, of the data. Mathematical expectation is a *weighted* average of possible outcomes, as the outcomes are multiplied by their probabilities. The more likely outcomes are weighted more than the less likely ones. A proportion is also an average of sorts. A baseball player who comes to bat five times in a game and gets two hits is said to have a batting average for the game of 2/5 = .400. This is the proportion of times at bat that he has hit safely but we can think of it as an average if we assign a value of 1 to each hit and a 0 to each out. His average for the game is the simple average of two 1s and three 0s. We can stretch the interpretation to mean he has averaged .400 hits per time at bat during the game.

Mathematical expectation can be used to predict future behavior. It requires a probability distribution of the possible outcomes. A reverse type of problem occurs when we have an outcome in the form of a sample taken from a population and are asked to estimate one or more parameters (an average, a proportion) associated with the population. *Inferential statistics* attempts to describe a population by making inferences based on a random sample taken from the population. It's a common process with which we are all familiar. Voter preference prior to an election is based on a randomly chosen sample from the population of voters. If 60 percent (a statistic) of the sample favors candidate X over candidate Y, we infer that the true proportion of all voters favoring candidate X is close to 60 percent (a parameter), plus or minus some margin of error. A sample's statistic is a *point estimate* of the associated population parameter. The mean (average) age of the population of voters can be similarly estimated by calculating the average age of a random sample of voters. For any given population, there exists no better form of parameter estimation than these point estimates.

Now consider the problem of estimating a set of k(possibly unrelated) parameters. As an example, assume the quality-control department of a widget manufacturer wants to estimate the percentage of defective widgets produced by each of its four independently operating assembly lines, A, B, C, and D. A random sample of 100 widgets is pulled from each assembly line and tested. Line A produces zero defective widgets, B produces one defective widget, C yields three defects, and D produces four. We can estimate the overall defect rates for each line as 0 percent for A, 1 percent for B, 3 percent for C, and 4 percent for line D.

Intuitively one might think this is our best estimate of the four defect rates. In 1961 Charles Stein and Willard James published the paradoxical result of there being better estimates, now known as James-Stein estimators, formed by combining the original four measured statistics. Why should the defect rate estimate for any one line be affected by the measured defect rates of the others if the lines operate independently? And how can there be a better estimate of the four parameters if each single measured statistic is the best point estimate of its associated parameter? This is the essence of Stein's paradox.

The James-Stein estimators are formed by *shrinking* the original four rates, call them r_A, r_B, r_C, and r_D, closer together, forming the improved estimates r_A^*, r_B^*, r_C^*, and r_D^*. To do so, first compute the overall average of the four measured rates, $\bar{r} = (r_A + r_B + r_C + r_C)/4$. Then, using a calculated shrinkage factor c (c is between

Assembly Line	Measured Defect Rate (r)	James-Stein Estimator (r^*)
A	0%	0.392%
B	1%	1.196%
C	3%	2.801%
D	4%	3.608%

Table 3.2. Rates r and r^*.

0 and 1), the James-Stein estimator for a given r is calculated to be $r^* = cr + (1-c)\bar{r}$. A shrinkage factor of $c = 0$ corresponds to all r values being shrunk down to the single James-Stein estimator $r^* = \bar{r}$. A shrinkage factor of $c = 1$ leaves each r value unchanged so that $r^* = r$ for each r. The formula for finding c is rather technical and the details are omitted here. For a detailed analysis of c's computation see [Efron and Morris 77] or [Ijiri and Leitch 80].

Getting back to our example involving the defect rate of the four assembly lines, and using $c = .804$, the original defect rates with their James-Stein estimators are given in Table 3.2.

In what sense are James-Stein estimators of parameters an improvement over the individually measured statistics? For a set of estimates, the overall error, or composite risk, is measured using the sum of squared errors (*SSE*). This is the sum of squares of the deviation of each estimate of a parameter from the true value. It has been shown (some examples to come) that the James-Stein estimators are expected to produce a smaller *SSE* than the observed statistics. Therefore, in our example, the r^*s are preferable to the rs as estimates of the true defect rates. The James-Stein estimators minimize the composite risk and, in doing so, may minimize the risk associated with each individual estimate. There is a chance

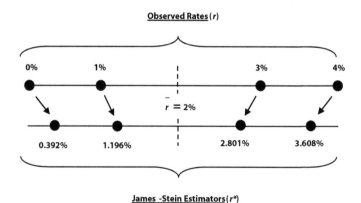

Figure 3.6. Shrinking toward $\bar{r} = 2\%$.

of some individual risks increasing when using the r's, but the overall *SSE* is expected to be less than that achieved using the measured rs. Think of things this way. Each measured r in our assembly line example is an estimate of an underlying true defect rate R. So, in general, the variability of the rs would exceed that of the true Rs. To better approximate the Rs, we should expect to shrink the set of rs to a common value, a reasonable choice for which would be $\bar{r}$. Figure 3.6 illustrates the shrinking process for our defect rate example.

Most surprising is that Stein's method of reducing the *SSE* using James-Stein estimators does not require that the parameters be related. The four assembly lines of our example operate independently. But, we could add to the mix a fifth statistic that approximates the prevalence of a flu virus during the month of January in Eureka, California. The five James-Stein estimators are expected to produce a smaller *SSE* than five individually measured rates. Surprising? Sure. The five populations (widgets from A, B, C, D and residents of Eureka) lack homogeneity, and this may contribute to undesirable increases in individual risk when using James-Stein estimators instead of directly measured (without shrinking) statistics. Nevertheless, the overall *SSE* is expected to be lower.

The paradox received worldwide attention when "Stein's Paradox in Statistics" by Bradley Efron and Carl Morris appeared in the May 1977 issue of *Scientific American* [Efron and Morris 77]. Efron and Morris estimate the batting averages of 18 major-league baseball players based on their first 45 times at bat during the 1970 baseball season. The James-Stein estimator is calculated for each of the players. The "true" batting average parameter for each player is taken to be the batting average maintained for each player during the remainder of the season. As expected, the James-Stein estimators of the players' true batting ability give a lower *SSE* than the original batting averages taken after the first 45 at bats. For 16 of the 18 players, the James-Stein estimator proves superior to the originally measured batting averages. Efron and Morris then go on to estimate a 19th parameter, the proportion of cars in Chicago that are foreign made. Though batting averages and imported cars are as unrelated as one can imagine, the procedure works equally well for the 19 parameters as it does for the 18 batters.

Phil Everson uses data from the 2006–2007 National Basketball Association (NBA) season to estimate the average points per game scored by each team as estimated by the average points per game scored by a team in its first ten games [Everson 07]. As expected, the James-Stein estimators are superior to the average points per game scored after the first ten games.

Agencies employ auditors to review accounts and transactions. The goal is to evaluate the agency's performance and make recommendations. Examining 100 percent of the available data is not cost efficient so samples are taken to estimate all relevant parameters. In a paper entitled "Stein's Paradox and Audit Sampling," Yuji Ijiri and Robert Leitch show how James-Stein estimators are used in the auditing context [Ijiri and Leitch 80]. Besides reducing both composite (*SSE*) and individual risk for most estimates, Stein's approach has the additional advantage of being purely objective. Other methods of audit sampling rely on subjective estimates of probabilities that may be difficult for the auditor to justify. Stein's method relies on samples taken from several populations in lieu of such estimates. It is far more objective, and the results may be easier to justify if the auditor is called upon to do so.

Appendix

Composite Sampling

Let x be the number of samples used to form the composite, let p be the proportion of individuals who are infected (p = prevalence), and let $q = 1 - p$. Using pooled samples, the expected number of tests is given by $E \approx q^x \cdot 1 + (1 - q^x)(x + 1) = x + 1 - xq^x$ and the expected number of tests per individual is

$$\frac{E}{x} \approx \frac{x + 1 - xq^x}{x} = 1 + \frac{1}{x} - q^x.$$

If $q = 0$ all individuals are infected and there would be $x + 1$ tests or $1 + 1/x$ tests per individual. If $q = 1$ then no individual is infected and there would be only one test (negative result), averaging $1/x$ tests per individual. Composite sampling would be preferential to individual testing only if the expected number of tests per individual is less than 1. To determine the minimum value of q (maximum value of p) for which this can happen, we set $1 + 1/x - q^x$ equal to 1 and set its derivative with respect to x equal to 0 yielding $q^x = 1/x$ and $1/x^2 + q^x \ln q = 0$. Solving this system of equations leads to $q = 1 / \sqrt[e]{e}$, $p = 1 - 1 / \sqrt[e]{e} \approx .31$, and $x = e$.

The Solvability of Tick-tack-toe, Checkers, and Chess

Tick-tack-toe (naughts and crosses) is a simple sequential game of perfect information. The players have complete recall of all previous moves as given on the nine-by-nine playing grid. In principle, an extensive form tree could be drawn. Assume Ron and Carla play with Ron going first. He has nine possible moves, each of which gives Carla eight possible moves, each of which gives Ron seven possible moves, and so on. An upper bound for the number of nodes in the bottom tier of the tree would therefore be 9! = 362,880. This includes illegal games that continue on after a player has won. It also includes equivalent positions of the game that differ only in the sense that one is obtained from the other by rotation. In any case, a large sheet of paper hundreds of feet wide by hundreds of feet long would be required to show the entire tree with font the size of the text you are reading now.

Despite the complexity of the tree, young children quickly learn how to play optimally. The game is fair in that both players are guaranteed no worse than a draw if they play optimally. The game soon becomes boring as its outcome is inevitable when played intelligently. When is the last time you played a challenging game of tick-tack-toe?

Checkers is far more complex. Its solvability has been known for some time; yet, having approximately 5×10^{20} positions that could arise during play, it is far more difficult to find optimal play than it is with tick-tack-toe. But it has been done! Jonathan Schaeffer of the Department of Computer Science at the University of Alberta spent 18 years on the project using dozens of computers running continuously. In 2007 Schaeffer published his results. His program, Chinook, plays perfect checkers and can't be beaten. If its opponent plays as well, the game will end in a draw. So, game over! Checkers is a fair game and the game will always end in a draw if both players make optimal moves at each step of the game.

Schaeffer's result is computer aided, and it is not practically possible to verify every step. So, there are skeptics, as there have been with other computer-aided proofs. The best-known example of such a proof is that of the four-color theorem, asserting that given any arbitrary map with countries, you need at most four different colors to color the countries so that no adjoining countries have the same color. A computational proof was given in 1976. If every step of the proof were to be written out and read for verification, no human could finish the task in a lifetime. To accept the proof, one must trust the unseen steps executed "inside the box."

Will chess be next? The number of possible positions is on the order of 10^{40}, far more than the number of checkers positions. Though known to be solvable, it may be well beyond our lifetimes before a solution similar to that for checkers is obtained.

Does the solvability of a sequential game suggest it may no longer hold our interest as a form of recreation? This may be the case with tick-tack-toe, but games like checkers and chess are so complex that perfect play may be beyond human reach, in which case the games remain challenging and fun to play. If there were no limits to human intelligence, then there would soon come the day that games like checkers, chess, and even go would be as boring as tick-tack-toe. Claude Shannon, considered by some as the "father of the information age," describes a chess game between two mental giants, Mr. A and Mr. B [Shannon 49]. They draw colors, survey the pieces, and then either

1. Mr. A says, "I resign" or
2. Mr. B says, "I resign" or
3. Mr. A says, "I offer a draw," and Mr. B replies, "I accept."

What fun!

A "Proof without Words" of Convergence

We want to show that

$$E = \sum_{n=1}^{\infty} 2^{-n} n = \frac{1}{2}(1) + \frac{1}{2^2}(2) + \frac{1}{2^3}(3) + \dots = 2.$$

We can think of the multiplication of two values as the area of a rectangle. The width of each column in Figure 3.7 is 1 and the figure extends infinitely far to the right. The horizontal lines are at heights of 1, 1/2, 1/4, 1/8, ... Starting from the top, the sum of the areas of the horizontal rows is 1/2 + 2/4 + 3/8 + ..., which is the series under consideration. This represents the total area of the infinitely extending figure, which we can also find by adding the areas of the vertical columns, starting at the far left. This sum is 1 + 1/2 + 1/4 + ..., shown to equal 1 + 1 = 2 in Chapter 2. The figure shows us that

$$\sum_{n=1}^{\infty} 2^{-n} n = 1 + \sum_{i=1}^{\infty} \frac{1}{2^i} = 2.$$

A picture's worth a thousand words!

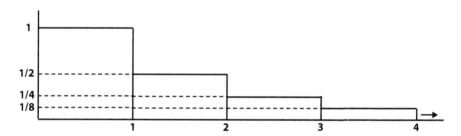

Figure 3.7. A "proof without words" of convergence.

Aversion Perversion

It is your aversion that hurts, nothing else.
—Hermann Hesse

$$\Psi + E = ?$$

The discussion of the previous chapter makes rational choice appear cut and dry. Our decisions, if based on the maximization of expectation, produce the best results in the long run and are rational as well for one shot (single-play) decisions. But why do we sometimes choose actions contrary to the maximization of expectation? Because we are human! Human decisions come with fears, emotions, subjective evaluations of risk, and other psychological components that compel us to behave not as suggested in the previous two chapters, but rather in ways from which we derive more comfort. Cold, calculated rational choices are often replaced by doing what feels right. The bumper sticker on the back of the 1960s VW bus reads, "If it feels good, do it!" And we do!

There are three well known aversions that affect our decisions. The most prominent is *loss aversion*. Obviously nobody plays to lose, but it is surprising how irrational we can be when avoiding losses. The second is *ambiguity aversion*, a special form of which is *deceit aversion*. Finally there is *inequity aversion*. All three are discussed in this chapter and to some degree contradict the notion of maximizing expectation as previously presented.

This chapter concludes with a discussion of subjective risk assessment. We are psychologically programmed to inflate some risks and deflate others, resulting in inaccurate subjective evaluations of expectation, which in turn leads to poor decisions.

Groundbreaking research in the field of behavioral economics was done by Daniel Kahneman and Amos Tversky, beginning with their seminal paper,

"Prospect Theory: An Analysis of Decision under Risk" [Kahneman and Tversky 79]. Their research shows, among other things, that humans are highly sensitive to losses, more so than to equivalent gains. To avoid losses, we exhibit *risk-seeking* behavior and to preserve gains we show *risk aversion*. Human behavior is therefore a function of how a problem is framed (in terms of potential losses or gains) as well as of the payoffs and associated probabilities. This may not be entirely rational, but research in this area is proving invaluable in such diverse areas as medicine (devising therapies), law (constructing legal arguments), business, and foreign affairs.

For his work in collaboration with Tversky, Kahneman was awarded the 2002 Nobel Prize in Economics. Tversky died in 1996.

Loss Aversion

Any loss is associated with varying degrees of stress, anxiety, disgust, pain, regret, disappointment, and possibly panic. Losing hurts! In fact, we feel twice as badly the pain of a loss than we feel the joy of an equivalent gain. Some professional athletes report hating to lose more than wanting to win. So, we tend to reject gambles where there is a 50/50 chance of winning or losing unless the amount that can be won is at least twice that of what could be lost. We might be willing to irrationally tolerate a high risk to avoid certain losses; otherwise, we are risk averse.

This idea is illustrated by the asymmetry of the graph in Figure 4.1 and a personal anecdote:

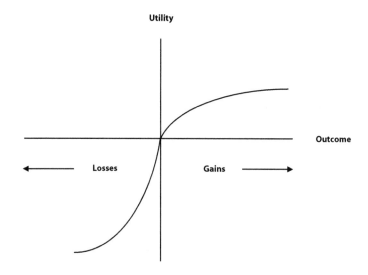

Figure 4.1. Loss aversion.

At El Camino College the school year begins in late August. Prior to the first day of class, faculty and staff are required to be on campus for a day of meetings and other administrative matters. My wife and I had made previous arrangements to be out of town that day, so I contacted my dean, asking if missing this non-teaching day would be a problem. I was told there would be no problem at all, except for the fact that I would have to take the day off as "personal business" and lose one day's salary. So, it was up to me. I discussed the matter with my wife, and we both agreed that losing several hundred dollars was a steep price to pay for a trip that could easily be rescheduled.

After giving the problem some thought, I was able to rationalize (*irrationalize*) going on the trip by framing the problem in terms of an unrealized gain as opposed to a loss. I reasoned, "I'm not losing anything. The money is not mine to lose since in order for me to lose it, it must be in my possession. Rather than losing, I'm failing to earn an additional few hundred dollars on my next paycheck."

In other words, when perceived as a loss, the pain was far more significant than when viewed in terms of an unrealized gain. The decision to take the trip was based, in large part, on the framing of the problem, as opposed to the actual payoff.

To illustrate the concept in terms of mathematical expectation and risk, imagine you are the emergency services director of a small town of 800 residents that has just experienced a natural disaster. You are presented with option set A and must quickly choose one of two possible actions, with the anticipated outcomes given.

A1. It is certain that 600 people will die.

A2. There is a 25 percent chance that nobody will die and a 75 percent chance that all 800 residents will die.

Which plan of action would you choose? Most people when faced with these two choices choose the risk-seeking option A2. Note that the expected number of lives saved is 200 for each course of action. But most are willing to gamble in the hope of avoiding a catastrophic loss.

Now imagine you are presented with option set B and must choose one of two courses of actions.

B1. It is certain that 200 people will be saved.

B2. There is a 25 percent chance that all 800 residents will be saved and a 75 percent chance nobody will be saved.

Which plan would you choose? Most people now choose the risk-averse option B1. Again note that both actions are associated with the expectation of saving 200 lives. But in scenario B, the problem is framed in terms of lives saved rather than lives lost and most chose the risk-averse option.

For the above two option sets, the four courses of action, A1, A2, B1, and B2 are all associated with the same expectation of saving 200 lives. So, our actions are best explained by our aversions to loss and risk as opposed to the actual expected outcomes. A slight change of numbers accentuates the irrationality of our actions. If, for option set A above, the percentages of 25 percent and 75 percent are respectively changed to 24 percent and 76 percent, then action A1 would yield the greater number of expected saved lives; most of us still choose option A2. Similarly, for option set B, if the percentages of 25 percent and 75 percent are respectively changed to 26 percent and 74 percent, action B2 would yield the greater number of saved lives; most of us still choose option B1.

Loss aversion explains *sunk-cost* dilemmas. A sunk cost is a previously incurred expenditure that can't be reversed. Therefore, it should not be taken into account when future decisions are made. Imagine your ten-year-old Chevy with over 200,000 miles on the engine goes into the shop and requires $800 of transmission repairs. You make the costly repair and two weeks later there is an engine problem requiring an additional $1,000 of repair costs. You anticipate additional expenses in the near future on a car with so many miles and, rationally, you might be better off cutting your losses, selling the car as is, and buying a new one. But what about the $800 you shelled out two weeks ago? Was it all for nothing? If you sell the car now without making the additional $1,000 of repairs, you will feel as if the $800 has been lost. And, if you spend the additional $1,000 then the $800 was not wasted. To avoid the feeling of loss, many of us would stick with the old clunker and repair it once again. Thinking rationally, one should ignore the $800 sunk costs as "water under the bridge." It should be irrelevant with respect to future decisions. But loss aversion compels us to act differently. This is precisely why many gamblers find it difficult to walk away from the table after losing a significant amount of money. In their mind, the loss isn't finalized until they walk away. In actuality, the money was lost when they lost the last bet and nothing they do now or later can change that fact.

Loss aversion has been seen in children as young as five years old and capuchin monkeys, suggesting that primate brains might be hardwired for such behavior. Yale University Economists have experimented with capuchins, giving them a choice between two games [Lakshminaryanan, Chen, and Santos 08]. In the first game the monkey was given one piece of apple and then a coin was flipped. Depending on the outcome, the monkey either kept the piece or received one additional piece. The expected number of pieces of apple the monkey will have at the end of the game is

$$E = \frac{1}{2}(1) + \frac{1}{2}(2) = \frac{3}{2} \text{ pieces.}$$

In the second game, the monkey is first given two pieces of apple and then, depending on the outcome of a coin toss, may lose one of the two pieces. The expected number of pieces at the end of the second game is obviously the same as the first. But, the vast majority of capuchins favored the first game, with the potential gain of one piece, as opposed to the second game, which involved a potential loss of one piece. Capuchins, like humans, hate to lose more than they like to win.

There may be physiological reasons for such behavior. Using magnetic resonance imagining (MRI) of the brain, University of California, Los Angeles researchers have shown a neural basis of loss aversion. Regions of the brain show an increased activity associated with potential gains and a decreased activity associated with potential losses [Tom, Fox, Trepel, and Poldrack 07]. Researchers are hopeful this may shed light on risk-seeking behavior and associated disorders such as substance abuse, gambling addiction, and antisocial personality disorders.

Those in the business of persuasion (attorneys, negotiators, counselors) should take heed. Arguments in favor of accepting something should focus on potential gains associated with the acceptance. For example, a surgeon should stress the survival rate when promoting a procedure having a 90 percent survival rate. Arguments in favor of rejection should stress the potential losses. Now the surgeon might stress the 10 percent mortality rate. These subtle psychological persuaders may make an otherwise weak argument convincing.

Ambiguity Aversion

As defined in Chapter 2, there are three types of probabilities than can be ranked in terms of how accurately they measure a likelihood of occurrence—theoretical probabilities, empirical probabilities, and subjective probabilities. If an urn contains 50 black balls and 50 white balls, then the theoretical probability of drawing a black ball is 50 percent. If an urn containing both colors of balls is placed before us and we do not know the exact number of each in it, we could select a small random sample, determine the proportion of black balls, and use this as an estimate of the theoretical probability. For example, if 50 percent of the balls in our small random sample are black, then the empirical probability (based on evidence) of drawing a black ball is 50 percent and we may infer that the theoretical (true) probability of drawing a black ball is 50 percent plus or minus some margin of error.

Subjective probabilities are the least reliable of the three probabilities specified. A subjective probability is a belief, or hunch, based on intuition without other significant evidence. We may suspect the probability of drawing a black ball from the urn is 50 percent based on a vague recollections of how the urn was filled or what some unreliable source may have told us about the process.

These three probabilities are measurable and may be quantified. In some cases, an uncertainty or risk may have no measure whatsoever, not even a subjective one. *Knightian uncertainty*, named after University of Chicago economist Frank Knight, refers to an uncertainty where we do not have enough information to specify a measure of likelihood. A closed urn contains an unknown number of black and white balls. The ratio of the two is not known. With no additional information being given, it would be impossible to assign a measure of likelihood to drawing a black ball on a single draw from the urn.

Rand Corporation economist Daniel Ellsberg, of Pentagon Papers fame, describes human behavior under Knightian uncertainty that appears to contradict principles of rational behavior [Ellsberg 61]. The Ellsberg paradox has various formulations; the two-urn version is given here.

Urn 1 contains 100 balls, 50 of which are black and 50 are white. Urn 2 contains 100 balls colored black and white in an unknown ratio. You are to draw one ball from the urn of your choice and place a bet on the ball's color. There are four options as to your choice of urn and bet:

1. Bet B1: You draw from urn 1 and bet on black.
2. Bet W1: You draw from urn 1 and bet on white.
3. Bet B2: You draw from urn 2 and bet on black.
4. Bet W2: You draw from urn 2 and bet on white.

You win $100 if the color specified is the color of the selected ball.

Do you have a preference between B1 and W1? Probably not. And it's unlikely you have a preference between B2 and W2. But what is your preference between B1 and B2? Most will choose B1. And what is your preference between W1 and W2? Most prefer W1. The preference of choosing the urn with the measurable risk (urn 1) over the urn with the unmeasurable risk (urn 2) is an example of *ambiguity aversion* and is reflected in the common adage, "better the devil you know than the one you don't."

The paradox arises when considering the expected payoffs associated with the above four bets. Denote the four expectations as E_{B1}, E_{W1}, E_{B2}, and E_{W2}. Since urn 1 contains 50 percent black balls and 50 percent white balls, it follows that $E_{B1} = E_{W1} = \$50$. The popular preference of B1 over B2 suggests that the player perceives E_{B2} as being less than $50, corresponding to the player's belief that the proportion of black balls in urn 2 is less than 50 percent. The player favoring B1 over B2 will most often also favor W1 over W2, suggesting the player believes the proportion of black balls in urn 2 is more than 50 percent. The player's beliefs with respect to the proportions are contradictory. (A similar contradiction would hold for the risk-seeking player who always chooses to bet on urn 2 rather than urn 1. We focus here on the more prevalent behavior of betting on urn 1.)

Ellsberg notes that the betting preferences are deliberate and do not change even when the contradiction is considered. It's also been noted that this type of aversion takes place in a comparative context, where the player has relatively more knowledge about one option than another. The aversion is significantly reduced in the absence of such a comparison [Fox and Tversky 95]. In the two urn example above, players would be less reluctant to place bets on urn 2 if urn 1 was not available as an alternative.

So, what causes the aversion to ambiguity?

Various explanations have been put forward. Betting on urn 1, of measurable risk, may be considered safer than urn 2, where the risk is unmeasurable. If the player were to play the game n times and consistently bet B1, then the player's expectation is that of winning $50n. Consistently betting B2 yields no measurable expectation since there is no information as to the contents of B2. Betting this way, the player's expectation could be anything between $0 and $100n, depending on the contents of urn 2. Playing conservatively to minimize the worst case scenario, the player may avoid the ambiguity of urn 2 and always bet on urn 1, despite the previously mentioned contradictions. (Interestingly, the unmeasurable risk of betting on urn 2 can be converted into a measurable risk by tossing a fair coin to determine on which urn 2 color to place a bet. If urn 2 bets are placed this way, then the risks associated with each of the two urns are equally measurable.)

The unknown contents of urn 2 may be associated with ignorance or incompetence. Player's may feel more comfortable placing bets where they have some knowledge than on matters of which they are totally ignorant. "Better the devil you know ...."

A third explanation of ambiguity aversion is associated with deceit. The player may associate the lack of information regarding urn 2 with intentional deception on the part of the experimenter. A player considering betting on a black ball chosen from urn 2 (B2) may fear being baited by the experimenter into placing the bet. If so, the player will perceive the proportion of black balls in urn 2 to be on the low end of its range (closer to 0 percent than 100 percent) and feel safer betting on a black ball chosen from urn 1 (B1), where the proportion of black balls is known to be 50 percent. The ambiguity of urn 2 triggers a fear of deception despite there being no deception on the part of the experimenter.

Ambiguity aversion contradicts our notions of mathematical expectation, as presented in the previous two chapters. But, should we write off such behavior as being totally irrational? Ellsberg has doubts [Ellsberg 61, p. 669]:

> Are they foolish? ... The mere fact that it conflicts with certain axioms of choice that at first glance appear reasonable does not seem to me to foreclose this question; empirical research, and even preliminary speculation, about the nature of actual or "successful" decision-making under uncertainty is still too young to give us confidence that these axioms are not abstracting away from vital considerations. It would seem incautious to rule peremptorily that the people in question should not allow their perception of ambiguity, their unease with their best estimates of probability, to influence their decision; or to assert that the manner in which they respond to it is against their long-run interest and that they would be in some sense better off if they should go against their deep-felt preferences. ... Indeed, it seems out of the question summarily to judge their behavior as irrational. ... They

act in conflict with the axioms deliberately, without apology, because it seems to them the sensible way to behave. Are they clearly mistaken?

Legal, financial, and other decisions involving risk are typically made in a comparative context where theoretical probabilities of potential outcomes are not known. Human behavior in such situations should be considered a function of the payoffs, probabilities, and ambiguities associated with the options. Our actions may appear irrational, in the sense that they do not maximize expectation. However, if we consider the axioms to be an abstraction of true human behavior, then we may need to revise the definition of rationality as it applies here.

Inequity Aversion

It's the *Golden Rule* and we're indoctrinated at an early age. Almost every religion has its version.

> Christianity: "You shall love your neighbor as you love yourself."
> Judaism: "What is hateful to you, do not do to your fellow man."
> Buddhism: "Just as I am so are they; just as they are so am I."
> Islam: "No one of you is a believer until he desires for his brother that which he desires for himself."

As children we are encouraged to "play fair," and playmates are quick to announce "No fair!" when rules are broken. Robin Hood and his Merry Men robbed from the rich and gave to the poor. Equality, balance, and fairness are associated with justice and generally viewed as desirable, whereas inequity and imbalance are often to be avoided. So, it may be no surprise that we are averse to inequities of both types—disadvantageous, where one has less than another, and advantageous, where one has more than another. What is surprising, however, is the magnitude of the aversion and the fact that it may cause otherwise rational individuals to fail to maximize expectation. Three well known experiments by social psychologists and behavioral economists show clear evidence of this behavior.

The Dictator Game

The *dictator game* is not a game in the true sense because only one of the two players makes a decision. The first player, the dictator, is given a fixed amount of money, say $100, and must divide this sum between himself and the second player. The dictator may choose to keep it all, give a portion of the money to the other player, or give all of it to the other player. The players do not know each other and may never meet. (Imagine the game played over the Internet.) Both players fully understand the nature of the game. To maximize the expected payoff, the dictator will keep $100 for himself and give nothing to the other player. Why do otherwise?

In actuality, the dictator often does otherwise, sometimes offering 50 percent ($50) to the other player. What compels a dictator to behave this way? It may be that we've evolved to the point where something in our DNA causes us to behave altruistically. Altruism, being good for society as a whole, may now be part of our makeup. It's more of an emotional decision than a rational one.

Some dictators report giving generous portions to the other player for fear of retribution, despite the guarantee of anonymity. The dictator may believe there remains a small chance of future interactions with the other player, in which case some sort of punishment for a selfish action will be delivered.

And finally there are higher authorities to respect. "What will the experimenter think of me if I keep the $100 for myself and give nothing to the other player?" Or, "How will God judge my selfish action?"

The dictator game is one of the simplest of all economic experiments, requiring no calculations or complex weighing of strategies on the part of the dictator. So, it is the ideal tool for measuring altruistic behavior in young children. A recent study involved children aged four, six, and nine years from British primary schools [Benenson, Pascoe, and Radmore 07]. The students were paired up and one child from each pair was allowed to choose ten stickers from a group of attractive stickers, the kind valued by children. Those that chose were then told they could keep all ten stickers if they chose to do so, or give a number away to another, anonymous student. All age groups exhibited significant degrees of altruism, with dictators giving away, on average, between two and four stickers. The degree of altruism shows a slight positive correlation with the age of the child and socioeconomic status. Four-year-olds donated, on average, between two and three stickers. Six-year-olds donated an average of three stickers. The average donation for the nine-year-olds was between three and four stickers. Within each age group, children of a higher socioeconomic status donated at a slightly higher rate than those of a lower socioeconomic status.

> Birds do it, bees do it,
> Even educated fleas do it.

Cole Porter's words refer to falling in love (or so say the lyrics); but it is also true that birds, bees, flees, and other species show altruistic behavior. For certain species of birds, breeding pairs receive help from unrelated birds in feeding and caring for offspring. Worker bees donate 100 percent of their lives to feeding, caring for, and protecting the queen. Adult fleas feed on a host, then excrete undigested blood upon which unrelated fleas feed.

African wild dogs show altruistic behavior when male pack members assist in raising the pups of a deceased female. Pups receive regurgitated food from unrelated adult members of the pack. And vervet monkeys issue alarm calls to warn other monkeys of predators. It's altruistic in the sense that it attracts attention to themselves, putting the monkey issuing the call in danger of being attacked.

The Ultimatum Game

Conceived in 1982 by the German economist Werner Güth, the *ultimatum game* has been used in hundreds of studies as a test of fairness. It is similar to the dictator game with the exception that the second (receiving) player is allowed to reject the offer and veto the deal if he believes his share is too low. If this occurs, the deal is off, the first player forfeits the $100, and neither player gets anything. As is the case with the dictator game, the players are mutually anonymous and both understand the rules of the game.

Acting rationally, the second player should be willing to accept anything, even an offer as low as $1. There would seem to be no rational reason for rejecting some amount of money in favor of absolutely nothing. Yet, rejection often occurs when the proposer's offer is low. The second player may get a sense of satisfaction by punishing the proposer for the unfair offer. The emotional satisfaction of righting a wrong may outweigh the potential material gain. Such vetoes would clearly make sense if the game were played repeatedly by the same pair of players, with the second player hoping to teach the proposer to "play fair." But to veto a one-time, single-play offer would be an emotional action contradicting the principle of maximizing expectation.

Capuchin monkeys have been shown to reject rewards if they believe more favorable rewards are being given to other monkeys. Researchers at Emory University in Atlanta report on experiments where monkeys may exchange tokens for cucumbers or more favored grapes. Monkeys exchanging the tokens for cucumbers were less willing to do so once they observed their partners exchanging for higher valued grapes [Brosnan and de Waal 03].

The Trust Game

Can we trust each other to play fair, even if such trust contradicts the maximization of expectation? The *trust game*, devised by economist Joyce Berg of the University of Iowa, suggests we can. In this game, two players are paired in mutual anonymity, as is the case with the dictator and ultimatum games. Both are initially given an equal sum of money, say $100, for agreeing to participate. The experimenter designates one player as the *trustor* and the other as the *trustee*. The two players are anonymous and may not communicate as the game is played. At stage one, the trustor is given the option of transferring a portion of the $100, call this amount T ($0 \leq T \leq 100$), to the trustee. Whatever is sent gets tripled before it reaches the trustee so that the trustee gets $3T$. At stage two, the trustee then has the opportunity to return back a portion of what has been received. Call the returned sum R ($0 \leq R \leq 3T$). Stage two is essentially the dictator game with the trustee acting as the dictator. The game is played only once, so there is no chance to communicate through repeated play. The game is diagrammed in Figure 4.2.

The trustor's decision as to how much to transfer to the trustee is based on, you guessed it, trust. The trustor hopes the trustee will reciprocate generously

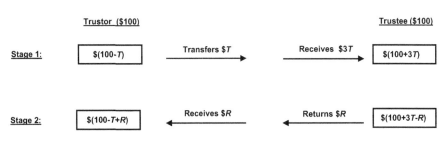

Figure 4.2. The trust game.

by returning a significant sum of money. If both trustor and trustee play in perfect trust they can double their respective accounts. For this to occur, the trustor would transfer the entire sum of $100 to the trustee, who would receive triple this amount, or $300, to be added to the trustee's initial account of $100. After the first transfer, the trustor would have $0 and the trustee would have $400. If the trustee decides to reciprocate by sending $200 back to the trustor, the game would conclude with each having doubled their original account. In general, equity is maintained whenever $100 - T + R = 100 + 3T - R$ or $R = 2T$.

Will rational players behave this way? They shouldn't, if they behave so as to maximize their respective gains. When the trustee receives the transfer of money from the trustor, there is no rational incentive to return any of it. The optimal action would be to return nothing ($R = 0$). With this understanding, the trustor should transfer nothing ($T = 0$) as there would be nothing expected in return and the money would be lost. This is precisely what traditional game theory predicts, that nothing would be transferred and nothing would be returned. Only irrational players would do otherwise.

In actuality, experiments show that trustors do exhibit a surprisingly high level of trust and trustees reciprocate generously. Roughly half of all trustees return more than was sent to them ($R > T$). Typically, perfect equity is not maintained; yet, there is evidence of inequity aversion on the part of the trustee shown by the reciprocal act of returning money to the trustor. And, trustors have faith that the trustees will, in fact, reciprocate.

Why does trust exist in such an environment? It may be that, as players, we are to be considered descendants of a long line of predecessors who have played the game over past centuries. Interactions in the form of social and financial contracts involve some degree of trust, a high level of which is beneficial to all. We have evolved to the point where trust is in our nature, even if it sacrifices the maximization of an immediate payoff. (The trust game is a special case of the prisoner's dilemma game where there is clear experimental evidence of the inadequacy of individual rationality. A full discussion of the prisoner's dilemma game is given in Chapter 8.)

Economists have shown there is a correlation between a nation's wealth and the level of trust shown by its citizens. When asked if people can be trusted, fewer than 10 percent of those from Brazil, the Philippines, and Uruguay said "yes." Forty-five percent of Americans claim to trust each other, and more than 60 percent of the respondents from Denmark, Norway, and other wealthier countries answered "yes" to the question. So, trust may be used as a predictor of the current and future wealth of a nation. Citizens of countries low in trust are reluctant to invest and engage in the transactions required to promote wealth. These countries may be associated with corrupt governments and no economic growth. High trust leads to investment, higher incomes, greater wealth, and even higher levels of trust. It's a self perpetuating loop.

Off-Target Subjective Probabilities

Aversions lead to irrational decisions. In the absence of such aversions, decisions based on maximizing expectations of the form $E = \Sigma xp$ would be rational. But, if the probabilities are inaccurate, then decisions based on E may yield suboptimal results.

Theoretical probabilities are 100 percent accurate and should not be problematic when used. In practice, such theoretical quantities may be unavailable and must be approximated by empirical probabilities, which may be reasonably good approximations of their respective theoretical probabilities. The least reliable estimates are subjective probabilities, often based on nothing more than intuition. These guestimates of the true probabilities are predictably off target, especially when it comes to highly visible events, often perceived as being more probable than they actually are. *Availability error* gets its name from our tendency to inflate the likelihood of those events that are easily brought to mind or that are readily available to our senses. Walking onto the floor of a large, Las Vegas casino, we immediately see and hear winners. The unmistakable sounds of coins falling onto the metal tray of a slot machine have since been replaced with flashing lights and sirens. Everyone on the floor is aware of the hundreds of winners. Nobody sees, hears, or even thinks about the thousands of losers. Consequently, players inflate their probability of winning and this works well for the casinos. Similarly, at a carnival or county fair, we notice those happy winners of large teddy bears and other stuffed animals as they proudly walk amongst us with their prizes. "If they can do it then why can't I? Look at all the winners!" We're not processing the significantly larger number of losers and the subjective probability of winning the prize becomes inflated.

Availability error occurs with vividly imaginable and unnatural causes of death such as air crashes, shark attacks, and homicide. We conjure up intense images that effectively magnify and distort our subjective estimations of their probability. In contrast, we underestimate less dramatic causes of death such as emphysema and stroke. Many of us believe homicide is a more likely cause of death than diabetes. In fact, diabetes claims twice as many lives. We may choose not to swim in the ocean, fearing a shark attack, while we relax on the beach while smoking a cigarette, a far riskier behavior than playing in the surf. We make poor decisions not always because we are irrational, but often because we calculate expectations using poorly estimated probabilities.

In 1994 a study was done in which 120 Stanford undergraduates were asked to estimate the probabilities of death from given possible causes [Tversky and Koehler 94]. Table 4.1 gives the means of their estimates as subjective probabilities. For comparison, empirical probabilities are given as close approximations of the true probabilities. Note that the probabilities of death by natural causes (heart disease, cancer, etc.) are underestimated and the probabilities of death by unnatural causes (accident, homicide, etc.) are overestimated. This is clear evidence of availability error because natural causes of death are less vivid in our minds than unnatural causes.

Further evidence of availability error was obtained when a second group of students was asked to estimate the probabilities of natural death and unnatural death, without subdividing the categories into specific causes as given in Table 4.1. The mean probability given for natural death was 58 percent, far lower than the sum of the subjective probabilities of death by different natural causes, given as 73 percent in Table 4.1. These students gave a mean group estimate for the probability of unnatural death as 32 percent, also lower than the sum total 53 percent given in the table. Both estimates by the second group of students are lower than the corresponding estimates of the first group because the specific causes were not given to the second group and the likelihood of their occurrence was less visible, and thus underestimated.

We should be less worried about flying than driving and not lose sleep over the possibility that our children may be kidnapped or that we will be killed by terrorists. We estimate some probabilities poorly, calculate off-target expectations, and often make poor decisions as a result.

Cause	Subjective Probability	Empirical Probability
Heart disease	22%	34%
Cancer	18%	23%
Other natural causes	33%	35%
All natural causes	73%	92%
Accident	32%	5%
Homicide	10%	1%
Other unnatural causes	11%	2%
All unnatural causes	53%	8%

Table 4.1. Estimates of probabilities of death.

For a sequence of random events clustered about a mean, an extreme outcome will most likely be followed by a less extreme one. The phenomenon known as *regression to the mean* explains why an extremely hot day is more often than not followed by a cooler one and a highly entertaining movie is followed by a less than spectacular sequel. Since, by definition, most events are not extreme, those that are tend to be followed by those that are not. This is due to chance alone.

A misunderstanding of this phenomenon may be associated with poor judgments with respect to the cause of an event and its likelihood of occurrence. Native American tribes believed ceremonial rain dances increased the chance of rain and a good harvest. In actuality, it is regression to the mean that would explain an eventual onset of showers following a significantly long dry spell.

Acupuncture, chiropractic treatment, and other forms of alternative medicine are sometimes credited by the patient as curing ills that appear nonresponsive to traditional treatment. But, back pain and severe headaches often subside by chance alone, returning the patient to a normal, less painful state. The well-known mantra *correlation does not imply causation* warns us not to believe a treatment cures a malady just because improvement follows the treatment.

Parents and teachers may believe punishment of bad behavior is a better teaching tool than rewarding good behavior because poor behavior, followed by punishment, often results in improvement, whereas good behavior is not as often sustained when followed by reward. Once again, this may be more a matter of regression to the mean than anything else. A very low test score may in all likelihood be followed by a better score on the next test because if the first test is extremely low, it is unlikely the student will score this poorly on the next test. If the student was punished or reprimanded, there would be a tendency to associate the improved score with the reprimand, rather than simply accepting it as regression to the mean. On the other hand, if the student did extremely well, possibly scoring 100 percent on a test, the score on the next exam could be no higher and in all probably may be somewhat lower. This would likely occur whether or not the student was rewarded for the high score on the first test.

In *How We Know What Isn't So: The Fallibility of Human Reason in Everyday Life*, Thomas Gilovich writes about an episode he witnessed on a trip to Israel [Gilovich 91, p. 28]:

A flurry of deaths by natural causes in the northern part of the country led to speculation about some new and unusual threat. It was not determined whether the increase in the number of deaths was within the normal fluctuation in the death rate that one can expect by chance. Instead, remedies of the problem were quickly put in place. In particular, a group of rabbis attributed the problem to the sacrilege of allowing women to attend funerals, formerly a forbidden practice. The remedy was a decree that subsequently barred women from funerals in the area. The decree was quickly enforced, and the rash of unusual deaths subsided—leaving one to wonder what the people in this area have concluded about the effectiveness of their remedy.

Complete ignorance of regression leads to highly inaccurate subjective probabilities via the *law of small numbers*. Bernoulli's law of large numbers states that the empirical probability of an event is expected to approach the true probability as the number of trials (sample size) increases. Empirical probabilities cluster about their expected value, the true probability of the event. But, if the sample size is small, the empirical probability may be excessively high or low, despite the fact that it is the best estimate based on the data at hand. The law of small numbers refers to using an empirical probability to estimate a true probability based on a small sample size, making the guestimate highly unreliable. It is not a law, but rather a mockery of the improper use of Bernoulli's law.

As an example, consider an urn that contains a large number of black and white balls. From a small sample of four balls, three are seen to be black. Our best estimate of the proportion of black balls in the urn is indeed three fourths; yet, we should not hold high confidence that the true probability of drawing a black ball in the future is three fourths. We need a much larger sample to be confident in our estimate.

Poor decisions, based on poor expectations, follow from such inaccurate estimates of probabilities. We go out to dinner at a new restaurant and have a bad experience. Our expectation is now lowered for the restaurant and we may never return, based on a sample of size one. Decisions based on first impressions are based on the law of small numbers and are often incorrect. If you visit London for the first time and it rains the first two days of your trip, do you conclude that it rains everyday in London?

A little knowledge is dangerous and regression to the mean can be incorrectly applied in the form of the *gambler's fallacy*. An even-money bet is to be placed on the outcome of a fair coin toss—heads you win, tails you lose. Prior to placing the bet, you observe the previous ten tosses all coming up tails. Knowing the coin is fair and that, in the long run, it will come up heads 50 percent of the time, you reason that it is now more likely to come up heads than tails, so as to make up for past history. Or, you may argue that the *law of averages* suggests there is now a greater probability of heads than tails. (The *law of averages* is a phrase, not well defined, that usually represents a misuse of the law of large numbers or regression to the mean.) You place your bet now, under the assumption that you are more likely to win than lose. In fact, by the law of large numbers, the empirical probability of heads will approach 1/2 as the number of tosses increases, but the coin is no more likely to come up heads now or in the future than it was on each of the preceding tosses. After all, the coin itself has no memory of what has transpired and in no way can it *choose* to make up for past history. Is regression to the mean being violated?

To understand, denote the empirical probability of heads as x/n, where x denotes the number of heads obtained after n tosses of the coin. The law of large numbers implies that x/n approaches 1/2 as n increases. But, this in no way implies that the difference between x and $n/2$ becomes arbitrarily small. In actuality, the difference generally tends to increase as n increases. Equivalently, the difference between the total number of heads and the total number of tails will probably widen. (For example, the fraction $(n^2 + n)/2n^2$ approaches $1/2$ as n increases because the fraction is equal to $\frac{1}{2} + \frac{1}{2n}$, which clearly has a limiting value of $1/2$. Yet, the difference between the numerator and one half the denominator is $n^2 + n - \frac{1}{2}(2n^2) = n$, which increases as n increases.)

The fallacy may occur with American roulette, where past spins of the wheel have produced a predominance of red numbers. A gambler watching the wheel understands that in the long run, the probability of the ball landing on a black (or red) number is slightly less than $1/2$. (Eighteen of the 38 numbered sectors are black, 18 are red, and two are green so $p(\text{black}) = p(\text{red}) = \frac{18}{38} = \frac{9}{19} \approx 47$ percent.) The gambler's fallacy occurs when the gambler incorrectly assumes that the wheel is now more likely to produce a black number than a red one, so as to ultimately achieve an equal proportion of black and red outcomes. But, assuming the wheel is not defective, the true probabilities remain the same on each spin of the wheel as the spins' outcomes are independent of previous outcomes. The expectation associated with betting $1 on black on the next spin of the wheel remains

$$E = \frac{18}{38}(\$1) + \frac{20}{38}(-\$1) = -\$\frac{1}{19} \approx -5\text{¢}.$$

A grand-scale gambler's fallacy occurred in 2004 when the number 53 failed to be drawn on 152 consecutive draws of a two-digit Italian lottery. Players, believing that the number 53 was due to be drawn, went into a ticket buying frenzy. Four deaths were blamed on the obsession, including one suicide where the woman blamed herself for spending the family's life savings betting on number 53. A man was arrested for beating his wife on a 53-related issue. Finally, after two years, the number was drawn and the obsession came to an end.

Another example of poor estimates of probability is the *hot-hand* or *hot-streak fallacy*. During sporting competitions, professionals and amateurs alike have experienced the hot-hand phenomenon, also known as *being in the zone*. It occurs when things are going well and the player has a sense of doing no wrong. The basketball player with the hot hand makes every shot, and the tennis player returns all shots with form and accuracy to win the point. The feeling may last for a few minutes or extend for several games (or matches). While in the zone, the player, teammates, and coaches all have increased expectations regarding the player's ability to score. Basketball players are encouraged to feed the player with the hot hand allowing that player to take shots that might not otherwise be taken. But, are the increased expectations justified? Is there a physical reality to being in the zone? Or, has intuition led us astray?

There's no denying that a player's frame of mind can affect his or her level of play. Feeling energized and inspired may lead to confident and aggressive play, producing good results. Alternatively, a player may lose confidence, become psyched out, and choke on key plays. But, in actuality, chance events are streakier than one might expect. If a fair coin is flipped 20 times, there is roughly a 50 percent chance of getting four heads in a row. (There is further discussion of our misconceptions about randomness in Chapter 10.) And, if chance alone is the reason behind the streaks and slumps, then a player should be expected to play no better and no worse during the streaks and slumps than during normal play. In fact, streaks and slumps are to be expected during normal play, but due to chance and not a change in the player's ability to execute plays.

In basketball, if the hot-hand phenomenon exists, then players should be more likely to make shots after making the previous shot than when missing the previous shot. And, there should be more streaks of shots made (and shots missed) than would normally occur by chance. Researchers from Stanford and Cornell used data obtained from the NBA Philadelphia 76ers during the 1980–1981 basketball season to investigate [McKean 81, p. 29]. This particular team was chosen for the study because they were the only NBA team that recorded the order of hits and misses for each player. The players estimated they were 25 percent more likely to make a shot after making the previous shot than after missing the previous shot. Statistics show that they were slightly more likely to make the shot after a miss than after a hit. In addition, it was found that streaks of hits and misses occurred no more often than chance alone would predict. The researchers found no evidence of there being a hot hand in basketball, based on the data of this study.

Similar studies about *hot-bat* hitting streaks in baseball show no evidence of the phenomenon; chance alone may well explain hitting streaks and slumps. Players, coaches, and most fans will undoubtedly disagree.

Figure 4.3. Fifty percent random pixilation.

An analogous geometric fallacy occurs when randomly occurring clusters appear and reasons other than chance are given as the explanation of their occurrence. Figure 4.3 is a 100 × 100 square grid of pixel cells, looking much like the QR codes (matrix barcodes) that are currently used for commercial tracking and in-store product labeling. Think of the random pattern of black and white pixels as being generated by 10,000 tosses of a fair coin—black if heads and white if tails. It's not surprising that approximately 50 percent of the pixels are black and 50 percent are white. What may be surprising is the degree of clustering that occurs by chance.

Some regions appear highly dense; others are relatively vacant. A careful inspection shows two 5 × 3 (15 pixel) rectangles, all pixels of which are black. For any given 5 × 3 rectangle, the probability of this occurring by chance alone is

$$\left(\frac{1}{2}\right)^{15} = \frac{1}{32768} \approx 0.003\%.$$

But, when one considers the vast number of patterns that can be formed from the 10,000 pixels, it should be no surprise that large black or white clusters exist.

During World War II, German V-1 bombs fell across London in what appeared to be clusters, suggesting to Londoners that some areas of their city were being targeted and were more likely to receive repeated strikes. Further analysis indicates the bombs landed randomly and the apparent "targets" were naturally occurring clusters such as those appearing in Figure 4.3. This form of the hot-hand fallacy is known as the "Texas Sharpshooter Effect," getting its name from someone randomly firing shots at the side of a barn and then afterwards painting the target around the most-clustered bullet holes.

Epidemiologists analyze geographic clusters of cancers, viruses, and other diseases to determine if environmental carcinogens or other nonrandom phenomena are causing the clusters. Law enforcement agencies check for geographical clusters of bank robberies, homicides, and other serious crimes and decide to increase surveillance and patrol in areas that appear to be at high risk. In some

Figure 4.4. Smirking human face.

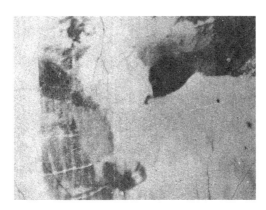

Figure 4.5. What do you see?

cases the efforts may be warranted, but often there is nothing more to it than chance.

We see what we want to see. Faces and other images appear in clouds, on the moon, and on satellite photos of the planets. *Pareidolia (apophenia, patternicity)* is the psychological phenomenon of seeing (or hearing) something of significance that in actuality is nothing more than a random occurrence. We are over-sensitized to perceive such patterns, especially those of the human face.

The curves and line in Figure 4.4 can be immediately recognized as a smirking human face. Why? No human face has ever come close to resembling such a crude stick-figure sketch. Our brains may be hardwired in such a way as to perceive significance when none exists in reality.

Much of this pseudo imagery is religious in nature, due to beliefs supported by faith. Some Christians flock to images of Jesus Christ or the Virgin Mary, which seem to regularly appear (as chance alone dictates) in such ordinary things as wood grain patterns in trees, moisture patterns on windows and buildings, and even food. It's difficult to understand how some of these sightings can be taken seriously. The image of Jesus has been seen in Marmite, tortillas, and Cheetos cheese snacks, named Cheesus by their discoverers.

Our predisposition to see faces and other everyday images hinders our ability to see the obvious.

Figure 4.5 is an actual, unretouched photograph of something recognizable by all of us. Most readers will not see it immediately. Hint: You must be in the right *mooooood* to see it clearly. No bull! It's an udderly amazing photo.

And the Envelope Please!

Less is more, more or less.
—Mies van der Rohe

A bird in the hand is worth two in the bush. Do you stick with a sure thing or risk it all and switch, hoping for something of greater value? It's a common dilemma inspiring the US television game show *Let's Make a Deal*. As mathematical recreations, envelope problems have been in circulation since the mid 1950s when Belgian mathematician Maurice Kraitchik published "The Paradox of the Neckties" in *Mathematical Recreations*, the English-language edition of *La Mathématique des Jeux* [Kraitchik 53, pp. 133–134]:

> Each of two persons claims to have the finer necktie. They call in a third person who must make a decision. The winner must give his necktie to the loser as consolation. Each of the contestants reasons as follows: "I know what my tie is worth. I may lose it, but I may also win a better one, so the game is to my advantage."
> How can the game be to the advantage of both?

Kraitchik gives a short discussion but offers no satisfactory resolution. Martin Gardner, perhaps the world's best-known author of mathematical recreations, further popularizes the problem as "The Wallet Game" in *Aha, Gotcha: Paradoxes to Puzzle and Delight*, using wallets rather than neckties in the formulation [Gardner 82, p. 106]. Whether it's neckties, wallets, or cold hard cash, the idea is the same. Stick with what you've got or exchange for another item of unknown value. We begin with the classic envelope problem, offer a solution, and then move on to variations with highly unexpected consequences.

91

The Classic Envelope Problem: Double or Half

A mathematical formulation of Kraitchik's necktie problem produces a *double or half* paradox.

Amounts of money are placed inside each of two identical envelopes. The envelopes are then sealed. All you know about the contents is that one envelope contains twice that of the other (see Figure 5.1). You randomly select one of the two envelopes, open it and note its contents. If, for example, you find it to contain $8, then you know the other, closed envelope must contain either $4 or $16. Reasoning that you could just as easily have chosen either envelope, you assume a probability of 1/2 for each of the two possibilities. Should you switch envelopes?

It would be rational to switch if the expected value of the contents of the other envelope exceeds your "bird in the hand" amount of $8. Calculating the expected value of the other envelope shows $E = \frac{1}{2}(\$4) + \frac{1}{2}(\$16) = \$10 > \$8$ and it appears you should switch. There's nothing paradoxical at this point. But then you reason more generally.

If your envelope contains $x, then

$$E = \frac{1}{2}(\$2x) + \frac{1}{2}\left(\$\frac{x}{2}\right) = \$\frac{5x}{4} > \$x$$

for all possible values of *x*. So, no matter what your envelope contains, you should switch envelopes to maximize your expectation. If the exchange is inevitable, why even bother opening your envelope? Exchange it straightaway. But after the exchange, and before opening your newly selected envelope, are you compelled to switch again? And then again?

One should expect a flaw in the above analysis and indeed there is. The assumption of a probability of 1/2 for each of the two possible values of the other envelope is incorrect. The correct probabilities are conditional and will depend on *x*, the contents of your chosen envelope, as well as the probability distribution of the envelope contents.

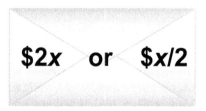

Your selected envelope   **Other sealed envelope**

Figure 5.1. Double or half.

Contents of Both Envelopes	Probability
{1,2}	1/3
{2,4}	1/3
{4,8}	1/3

Table 5.1. Probability distribution for {1,2}, {2,4}, and {4,8}.

For example, assume there are three equiprobable possibilities for the contents of the envelopes: {1,2}, {2,4}, and {4,8}. So, the probability of each is 1/3. The probability distribution is given in Table 5.1.

If you open your selected envelope and it contains $4, then there is an equal chance the other envelope contains either $2 or $8. The expected value of the other envelope's contents is $E = \frac{1}{2}(\$2) + \frac{1}{2}(\$8) = \$5$. Since this is greater than $4, you should switch envelopes. Each possible value of x should be individually considered and the results are summarized in Table 5.2.

Contents of Your Envelope (x)	Expected Value of the Other Envelope	Switch?
1	2	Yes
2	5/2	Yes
4	5	Yes
8	4	No

Table 5.2. Expected value of the other envelope.

So, switch for x = 1, 2, or 4 and keep your envelope if x = 8.

For the same possible contents, let's look at a slightly different probability distribution, in Table 5.3.

An analysis similar to the preceding example shows when to switch; see Table 5.4.

In this case, you should switch only if your chosen envelope contains $1. For all other values of x it is best not to switch.

So, the decision to switch is determined by the contents of the selected envelope as well as the probabilities associated with all possible contents. In a paper

Contents of Both Envelopes	Probability
{1,2}	2/3
{2,4}	1/4
{4,8}	1/12

Table 5.3. A non-equiprobable probability distribution.

Contents of Your Envelope (x)	Expected Value of the Other Envelope	Switch?
1	2	Yes
2	20/11	No
4	7/2	No
8	4	No

Table 5.4. Expected value of the other envelope.

entitled "The Box Problem: To Switch or Not to Switch," Steven Brams and D. Marc Kilgour give a specific criterion for switching, which they refer to as the "General Exchange Condition" [Brams and Kilgour 95]. The condition and its proof are given in this chapter's appendix.

Do envelope problems exist for which switching is rational for *all* values of x, as was falsely suggested by the classic envelope problem? The unexpected answer is yes, as will be shown.

The St. Petersburg Envelope Problem

The classic envelope problem is not a paradox if proper attention is given to the probability distributions of the envelopes' contents. But, a slight variation kicks it up a notch and we're no longer in Kansas. An amount of money is placed in one envelope by a St. Petersburg procedure as described in Chapter 3. That is, a fair coin is tossed repeatedly until heads first appears. If this happens at the nth toss, then a sum of $\$2^n$ is placed into this envelope. The envelope is then closed and sealed. A similar procedure is independently performed to place a dollar amount (a power of two) into the other envelope. It too is closed and sealed. You randomly choose one envelope, open it, and observe its contents to be $\$x$ (a power of 2). If given the opportunity, should you forfeit the contents of your chosen envelope in exchange for the unknown contents of the other, sealed envelope?

Would you opt to exchange what you have, a known finite amount of money, for an undetermined amount of infinite expected value? (See Figure 5.2.)

Your selected envelope   **Other sealed envelope**

Figure 5.2. St. Petersburg envelope problem.

For example, if your envelope contains $2^3 = 8$, it would make sense to exchange the $8 for an amount of infinite expectation. In the long run, you're guaranteed to be better off; so, for a one-time play, the rational choice would be for you to switch. In fact, it would make sense to exchange any finite amount of money in exchange for an undisclosed amount of money if the expected value of the undisclosed amount exceeds the finite amount. And, this will always be the case if the undisclosed amount has infinite expected value. By this line of reasoning, you should always switch.

Strange! If one should always switch after observing the contents of the chosen envelope, then why open the envelope? No additional significant information will be gained in opening the chosen envelope, and one might as well switch immediately and get it over with. Switching appears to be the rational choice in all cases, and there is no generally accepted resolution to the paradox.

What would happen if you choose one envelope and your friend chooses the other? Should both of you feel rationally compelled to switch and exchange, prior to opening your respective envelopes? How much would you be willing to pay your friend for the right to exchange envelopes prior to opening?

Why not use Daniel Bernoulli's approach to the St. Petersburg problem as given in Chapter 3? Let the player's utility of the payoff be the base 2 logarithm of the envelope's content. This temporarily resolves the envelope problem because the expected utility of the other, unopened envelope is now 2 and the paradox disappears. But, as we saw in Chapter 3, if the problem is modified so that the contents of each envelope is 2^{2^n} instead of 2^n, then the expected utility of the other envelope is once again infinite and the paradox reappears.

The "Powers of Three" Envelope Problem

A schoolboy is reported to have described infinity as "a place where things happen that don't." The St. Petersburg envelope problem involves the comparison of a finite amount of money, x dollars, with a closed envelope of infinite expected value. So, we might expect paradoxes. David Gale, professor emeritus of mathematics at the University of California, Berkeley, and former associate editor of *The Mathematical Intelligencer*, gives an envelope problem where the observed value x and the expected value (assuming x has been noted) of the other unopened envelope are both finite. This problem and others are found in his book, *Tracking the Automatic Ant and Other Mathematical Explorations* [Gale 98].

As before, you are presented with two closed envelopes, each containing a different sum of money. The money was placed in the envelopes by the following scheme. A fair coin was tossed repeatedly until heads first appeared. If this occurred on the nth toss, then 3^n was placed in one envelope and 3^{n+1} was placed in the other (see Figure 5.3). As before, you choose an envelope, open it, and observe the contents. Based on your observation, should you switch? This is similar to the classic envelope problem; instead of *double or half* it's *triple or third*.

 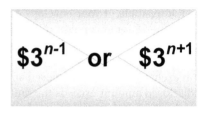

Your selected envelope   **Other sealed envelope**

Figure 5.3. Powers-of-three envelope problem.

You reason as follows:

If your envelope contains $3 (the minimum), you should definitely switch as the other envelope is sure to contain $9. If your envelope contains $9 then the other envelope is sure to contain either $3 or $27, with $3 being twice as likely as $27. Why? For the other envelope to contain $3, heads must occur on the first toss. This occurs (unconditionally) with probability 1/2. For the other envelope to contain $27, heads must not occur until the second toss and this occurs (unconditionally) with probability $\frac{1}{2} \times \frac{1}{2} = \frac{1}{4}$. So, the first outcome is twice as likely as the second, requiring the respective conditional probabilities to be 2/3 and 1/3, given that one of the two outcomes must have occurred. The expected value of the other envelope is therefore $E = \frac{2}{3}(\$3) + \frac{1}{3}(\$27) = \$11 > \$9$ and you should switch envelopes.

In general, if your envelope contains 3^n ($n > 1$) then the expected value of the other envelope is $E = \frac{2}{3}(\$3^{n-1}) + \frac{1}{3}(\$3^{n+1}) = \frac{11}{9}(\$3^n) > \$3^n$. So, as was the case with the St. Petersburg envelope problem, you are rationally compelled to switch regardless of the contents of your envelope. So, why open it? Just switch? But should you feel compelled to switch ad infinitum? Here we go again!

For both the St. Petersburg envelope problem and the powers-of-three envelope problem, the expected value of the envelopes are infinite, prior to the observation of the contents of the chosen envelope. Some offer these paradoxes as arguments that infinite expectation has no meaningful interpretation.

Are there envelope problems for which one should always stick and never switch? It's easy to envision such examples if we allow negative values for contents of the envelopes. Instead of dollars placed in the envelopes, imagine bills, debts, or similar penalties at specified amounts. In the preceding example, imagine the envelopes' contents to be penalties assessed once the envelopes are opened. If such penalties are represented by negative numbers, then the expected value of the other envelope is given by

$$E = \frac{2}{3}(-\$3^{n-1}) + \frac{1}{3}(-\$3^{n+1}) = \frac{11}{9}(-\$3^n) < -\$3^n$$

and it would never be rational to switch envelopes.

For both the classic envelope problem (double or half) and the powers-of-three envelope problem (triple or third), where the contents of the envelopes are assumed positive, no distributions exist for which sticking is the rational choice for all observed x. The proof is given in this chapter's appendix.

Blackwell's Bet

Two closed envelopes contain unequal sums of money (see Figure 5.4). To simplify things, assume the contents are positive integers. The probability distributions are unknown. You randomly choose one of the envelopes, open it, and observe the contents to be x dollars. You must now predict whether the other, closed envelope contains a sum of money more, or less, than x dollars. Since the distributions are unknown, you know nothing about the contents of the other envelope and have a 50 percent chance of guessing correctly. Right?

Unexpectedly, there is something you can do, short of opening the other envelope, to give yourself a better than even chance of getting it right. The simple method is a modification of work by David Blackwell [Blackwell 51, pp. 393–399]. By any means whatsoever, select a random positive integer. A simple way to do this would be to flip a coin repeatedly and let d be the number of flips required for heads to first appear. (If $d = x$ then repeat the process until this is not the case.) If $d > x$ guess higher and if $d < x$ guess lower. Simply put, d points in the direction of the unknown y. You will guess correctly more than 50 percent of the time simply because d points correctly more than 50 percent of the time!

Why is this so? If d is to be chosen randomly and independently of the envelopes, how could it possibly point in the right direction most of the time? Are we to believe that a randomly chosen number can assist in this difficult prediction? We should!

Think of it this way. If d falls between x and y then your prediction (as indicated by d) is guaranteed to be correct. Assume this occurs with probability p. If d falls less than both x and y, then your prediction will be correct only in the event your chosen number x is the larger of the two. There is a 50 percent chance of this. Similarly, if d is greater than both numbers, your prediction will be correct only if your chosen number is the smaller of the two. This occurs with

Your selected envelope

Other sealed envelope

Figure 5.4. Blackwell's Bet.

a 50 percent probability as well. So, your overall probability of being correct is $p + (1 - p)(\frac{1}{2}) = \frac{1}{2} + \frac{p}{2} > \frac{1}{2}$. The odds are in your favor of making a correct prediction. It's unexpected and ironic that an unrelated random variable can be used to predict that which appears to be completely unpredictable. This is reminiscent of Stein's paradox (Chapter 3) where statistical estimations of parameters can be improved by considering additional statistics that may be unrelated to the parameter being estimated.

The technique works for similar situations, as long as the values can be quantified, or at least ordered. Imagine you are at a winery, trying to decide which of two bottles of wine to purchase. The winemaker will only allow you to taste one and you have no means of making a direct comparison. You taste one of the two and must now decide whether to purchase this wine or take a chance on the other, which might taste better or worse than the wine you sampled. What do you do? You tell the winemaker you will return the next day to purchase one of the two wines. That evening at dinner you order a glass of wine, unrelated to the two wines in the shop. It tastes better than the wine you sampled and therefore *points* to the other, untasted wine at the winery. You return the next day and purchase the untasted wine, knowing that for you it is probably the better tasting of the two wines.

If we can predict if an unopened envelope contains more (or less) than what we presently have, and the tastiness of an untasted wine, then why not employ the same technique for predicting other future events? If we know the value of today's closing Dow Jones Industrial index, can we predict, with a probability greater than 1/2, if tomorrow's closing value will be higher or lower than it is today? If only it were so!

J. Laurie Snell and Robert Vanderbei offer this version of Blackwell's technique [Snell 95, p. 359]. Let a be an unknown integer, to which we add the random variable b of value 1 or −1, each with probability 1/2. We observe the sum c and are asked to guess the value of b. Since $a = c - b$, this is equivalent to guessing the value of a. Using the Blackwell procedure, we can guess correctly more often than not. By any means, choose a random real number d. Presumably $d \neq c$; otherwise, choose again. If $d < c$ guess $a = c - 1$ ($b = 1$); otherwise, guess $a = c + 1$ ($b = -1$). You will be correct more often than not.

The Snell–Vanderbei example can be applied in a surprising way to random walks. We use the outcome of a coin toss to determine how a marker advances on a number line (see Figure 5.5). Initially place the marker at position n, where n denotes a given integer. The coin is tossed and if the outcome is heads, advance to position $n + 1$; otherwise, fall back to position $n - 1$. Repeating this procedure generates a random walk among the set of integers, similar to that of a drunken individual who, at each step, is equally likely to step forward as fall back. For a given n, the random walk sequence might appear something like $n, n - 1, n - 2, n - 3, n - 2, n - 1, n, n + 1, \ldots$. Assume that we observe the present position of a marker moving as described. Say it is at position 10. We have no idea as to the initial position of

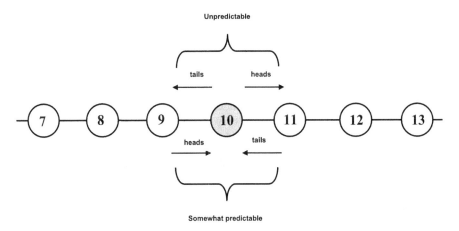

Figure 5.5. Random walk.

the marker nor do we know how many steps have transpired. If asked to predict the next position, we can do so with probability no better (or worse) than 1/2 as this is equivalent to guessing the outcome of the very next toss of the coin. We can do better with respect to guessing the position of the marker one step before it reached its present position. We directly apply the Snell–Vanderbei technique, and we can guess correctly more often than not. This is equivalent to correctly guessing the outcome of the preceding toss more often than not.

Was it Niels Bohr or Yogi Berra who once remarked, "Prediction is very difficult, especially about the future"? Though humorous, the statement suggests something remarkable about random walks. Given knowledge of the marker's present position, we can guess the previous position and outcome of the previous toss with greater success than we can for the following position and toss. Strange but true!

The Monty Hall Problem

The *stick or switch* dilemma is the heart of a once popular US television game show, *Let's Make a Deal*, which first aired in 1963 and ran for twenty-seven years. The show has been produced worldwide and an updated version aired in the US in 2009. A contestant is presented with three closed doors (curtains, boxes), behind which are prizes of variable worth. After the contestant selects one of the three doors, the option is given to stick with the prize (sometimes revealed) behind the chosen door or switch to an unknown prize behind one of the other two doors. The show has been generally recognized as one of the greatest TV game shows of all time.

A widely discussed mathematical problem of conditional probability is now known as "The Monty Hall problem," named after the show's host. The problem first appeared when Marilyn vos Savant, a columnist for *Parade Magazine*, published the following question posed by one of her readers in the magazine's column *Ask Marilyn* [vos Savant 96, p. 6]:

> Suppose you're on a game show, and you're given a choice of three doors. Behind one door is a car; behind the others, goats. You pick a door, say, number 1, and the host, who knows what's behind the doors, opens another door, say number 3, which has a goat. He then says to you, "Do you want to switch to door number 2?" Is it to your advantage to switch your choice of doors?

Her column regularly posed all sorts of questions, many from mathematics and general science. She was listed at that time in the *Guinness Book of Records* for having the world's highest IQ of 228 and therefore had some credibility when it came to answering questions of this sort. Her answer to the posed question was "Yes," suggesting the contestant would be advised to switch to door number 2. Over the years the problem has received more than its 15 minutes of due fame.

Her answer is correct and its mathematical justification is relatively simple. What's most surprising is the amount of attention the problem received and the number of individuals, many highly educated, who disagreed with her answer. She received over 10,000 letters in response to her answer, most in disagreement and highly critical. The following are typical [vos Savant 96, pp. 5–16]:

> Since you seem to enjoy coming straight to the point, I'll do the same. You blew it! Let me explain. If one door is shown to be a loser, that information changes the probability of either remaining choice, neither of which has any reason to be more likely, to 1/2. As a professional mathematician, I'm very concerned with the general public's lack of mathematical skills. Please help by confessing your error and in the future being more careful.
>
> —R. S., Ph.D.
> George Mason University

> You blew it, and you blew it big! Since you seem to have difficulty grasping the basic principle at work here, I'll explain. After the host reveals a goat, you now have a one-in-two chance of being correct. Whether you change your selection or not, the odds are the same. There is enough mathematical illiteracy in this country, and we don't need the world's highest IQ propagating more. Shame!
>
> —S. S., Ph.D.
> University of Florida

> Your answer to the question is in error. But if it is any consolation, many of my academic colleagues have also been stumped by this problem.
>
> —B. P., Ph.D.
> California Faculty Association

You're in error, but Albert Einstein earned a dearer place in the hearts of people after he admitted his errors.

—F. R., Ph.D.
University of Michigan

I have been a faithful reader of your column, and I have not, until now, had any reason to doubt you. However, in this matter (for which I do have expertise), your answer is clearly at odds with the truth.

—J. R., Ph.D.
Millikin University

May I suggest that you obtain and refer to a standard textbook on probability before you try to answer a question of this type again?

—C. R., Ph.D.
University of Florida

I am sure you will receive many letters on this topic from high school and college students. Perhaps you should keep a few addresses for help with future columns.

—W. S., Ph.D.
Georgia State University

You are utterly incorrect about the game show question, and I hope this controversy will call some public attention to the serious national crisis in mathematical education. If you can admit your error, you will have contributed constructively towards the solution of a deplorable situation. How many irate mathematicians are needed to get you to change your mind?

—R. B., Ph.D.
Georgetown University

I am in shock that after being corrected by at least three mathematicians, you still do not see your mistake.

—K. F.
Dickinson State University

Maybe women look at math problems differently than men.

—D. E.
Sunriver, Oregon

You are the goat!

—G. C.
Western State College

You made a mistake, but look at the positive side. If all those Ph.D.'s were wrong, the country would be in some very serious trouble.

—E. H., Ph.D.
US Army Research Institute

Savant patiently replies,

Gasp! If this controversy continues, even the postman won't be able to fit into the mailroom. I'm receiving thousands of letters, nearly all insisting that I'm wrong, including the Deputy Director of the Center for Defense Information and a Research Mathematical Statistician from the National Institutes of Health! Of the letters from the general public, 92 percent are against my answer, and of the letters from universities, 65 percent are against my answer. Overall, nine out of ten readers completely disagree with my reply.

Now we're receiving far more mail, and even newspaper columnists are joining in the fray! The day after the second column appeared, lights started flashing here at the magazine. Telephone calls poured into the switchboard, fax machines churned out copy, and the mailroom began to sink under its own weight. Incredulous at the response, we read wild accusations of intellectual irresponsibility, and, as the days went by, we were even more incredulous to read embarrassed retractions from some of those same people!

So let's look at it again, ...

The highly popular television show *NUMB3RS* featured the problem in its last show of the 2004–2005 television season. In the movie *21*, Kevin Spacey plays an MIT professor of mathematics and uses the Monty Hall Problem to challenge his students. A 1991 Sunday issue of the *New York Times* gave the problem front page attention. The problem is also discussed in Mark Haddon's best seller, *The Curious Incident of the Dog in the Night-Time*. The list goes on.

An equivalent (envelope) version is presented here with the mathematical justification for Savant's correct answer.

The host displays three identically appearing sealed envelopes. You know for a fact that one envelope contains $300 and the other two are empty. You randomly select one of the envelopes, but may not open it. Its contents ($0 or $300) are yours to keep, unless you decide to switch envelopes. To make your decision easier, the host opens an envelope that he knows beforehand to be empty. You verify that it is, indeed, empty. (See Figure 5.6.)

You are now given the choice of keeping your envelope or switching.

Your selected envelope   Other sealed envelope   Opened envelope

Figure 5.6. The Monty Hall problem.

Which of the following is correct?

(a) Keep the chosen envelope as its expected value is greater than that of the remaining sealed envelope.
(b) Switch to the second, sealed envelope as its expected value is greater than that of your chosen envelope.
(c) It doesn't matter whether you switch or not. Both your envelope and the other sealed envelope have an equal probability of containing the $300.

If seeing this problem for the first time, you might choose (c), that it would make no difference whether or not you switch. You're not alone in answering incorrectly. You might reason that you're holding one of two closed envelopes, one of which contains $300, giving you an equal chance of having the money or not. Then there would be a probability of 1/2 that your envelope contains $300 and a probability of 1/2 that it doesn't. The expected value of your envelope is $150 and the expected value of the other is also $150, suggesting that switching wouldn't make a difference one way or the other.

A correct analysis shows otherwise. Before the other sealed envelope is opened, your envelope contained $300 with probability 1/3 and its expected value was $100. When the host opens the empty envelope and shows it to you as empty, no additional information is being provided. You knew beforehand that an empty envelope was to be opened. (Presumably, the host would have no reason to reveal the envelope containing the money, had you not chosen it originally.) The probability remains 1/3 that your envelope contains $300, in which case the probability of the unselected sealed envelope containing $300 is 2/3. So, the expected value of your envelope is $100 and the expected value of the remaining, unselected envelope is $200. Switching doubles your expectation and you should switch. Despite the volume of criticism, Savant got it right.

Not convinced? Try thinking of it this way. A stranger offers you $300 if you can correctly guess her birthday. You know you've got virtually no chance of guessing correctly; but, with nothing to lose, you give it a shot and guess May 1 while walking away in resignation. The stranger tells you she'll make it easier by eliminating all but two possible dates, your choice of May 1 or September 18. Would you be willing to switch? Hopefully so, as the stranger's birthday is more likely to be September 18.

Singer/songwriters Jimmy Buffet and Steve Goodman memorialize Monty Hall and *Let's Make a Deal* in song:

> Oh, Monty, Monty, Monty, I am walkin' down your hall
> Got beat, lost my seat, but I'm not a man to crawl
> Though I didn't get rich, you son of a bitch
> I'll be back just wait and see
> 'Cause my whole world lies waiting behind door number three
> Yes, my whole world lies waiting behind door number three

Win-Win

The problem below is technically not an envelope problem, though it can be formulated as such. It fits in with this discussion and the analysis is similar. The presentation here is a modification of one given by David Gale [Gale 98, p. 8].

A casino posts a positive integer n at a gaming table. You, the player, will then toss a fair coin repeatedly until heads first appears (see Figure 5.7). If $n - 1$ tosses are required, you pay the casino $\$8^{n-1}$. If $n + 1$ tosses are required, the casino pays you $\$8^n$. (See Figure 5.7.) Otherwise, no money is exchanged.

After each play, the posted number is reset by the casino and you may play again, or allow a new player to approach the table. Only one player can participate at a time.

Assume the posted number is 3. Is the game in your favor? Is your expectation positive? A simple calculation shows this is so:

$$E = \frac{1}{4}(-\$8^2) + \frac{1}{16}(\$8^3) = \$16 > \$0.$$

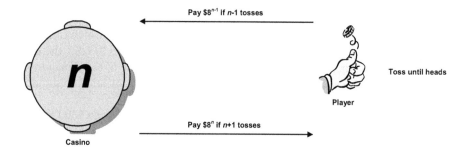

Figure 5.7. Win-win.

The positive expectation indicates the game is in your favor. In fact, this is so for any posted number n. Note that if the casino posts $n = 1$, then $E = \frac{1}{4}(\$8^1) = \$2 > \$0$. If $n > 1$ then $E = \frac{1}{2^{n-1}}(-\$8^{n-1}) + \frac{1}{2^{n+1}}(\$8^n) = \$2^{2n-2} > \$0$. So, no matter what number is posted at the table, you play with a positive expectation. Players believe they have a mathematical edge over the house, and the game's popularity increases.

The casino reasons differently. The casino secretly decides which number to post by using the same coin-tossing procedure as used by the player. Like the player, the casino tosses a fair coin repeatedly until heads first appears. If it appears on the nth toss, then the number n is posted at the table. Under these conditions, the game is symmetrical. As Gayle succinctly puts it, "the n tosser pays the $n + 1$ tosser $\$8^n$." The casino symmetrically reverses the player's reasoning and is convinced the game will generate a profit.

To accommodate the gamblers' increased demand, the casino decides to add more such tables to the casino floor, each posting its number as described. Gamblers with a low-risk tolerance play the tables with relatively small posted numbers, and those willing to risk more seek out the tables with larger posted numbers. Both the casino and the gamblers believe the game is in their favor.

Is it a win-win situation? Impossible! The explanation is given in the appendix.

Appendix

Exchange Conditions

Brams and Kilgour give a quick-and-easy rule to determine when to switch in the classic envelope problem and the powers-of-three envelope problem. The problems are similar in the sense that the first is *double or half* and the second is *triple or third*.

For the classic envelope problem (double or half), assume you open one envelope and observe its contents. We can simplify the notation without altering the result by assuming the contents to be a power of 2, say $\$2^n$. Then possible values for the other envelope are $\$2^{n-1}$ or $\$2^{n+1}$. The other envelope contains either half or double what you presently see in your envelope. So, there are two possibilities for the contents of both envelopes. Denote the respective unconditional probabilities as p_- and p_+, where the subscript reflects the relative value of the other envelope; see Table 5.5. These are the same probabilities given in similar tables throughout this chapter.

Contents of Both Envelopes	Probability
$\{2^{n-1}, 2^n\}$	p_-
$\{2^n, 2^{n+1}\}$	p_+

Table 5.5. Exchange condition.

A switch is beneficial if the expected value E, of the other envelope, exceeds 2^n. In this case,

$$E = \frac{p_-}{p_- + p_+}(2^{n-1}) + \frac{p_+}{p_- + p_+}(2^{n+1}) > 2^n \Rightarrow \frac{p_-}{p_- + p_+} + \frac{4p_+}{p_- + p_+} > 2 \Rightarrow p_- < 2p_+.$$

The inequality $p_- < 2p_+$ is the condition required for switching. Checking the probability distributions given in this chapter, it is verified that switching is desirable if and only if this condition holds.

For the powers-of- three envelope problem (triple or third), a similar analysis shows a switch is warranted if and only if $p_- < 3p_+$. For any observed contents of the form 3^n, $p_- = \frac{1}{2^{n-1}}$ and $p_+ = \frac{1}{2^n}$. So, we verify that $p_- = 2p_+ < 3p_+$ for all n, and one should always switch envelopes, regardless of the contents of the chosen envelope.

If we're restricted to positive contents, then a simple argument shows that, in problems similar to these, a switch is required for at least one value of the observed contents x. That is, there are no distributions where one should always stick.

To prove this for the classic envelope problem, begin by assuming the contrary. Then $p_- \geq 2p_+$ for all possible observed values x. There can be no minimum possible observed value as this would surely require one to switch. So, the full probability distribution given in table form has infinitely many rows with no uppermost row. The above inequality is impossible as it would require that infinitely many probabilities exceed 1, contradicting the definition of probability. So, it must be that switching is required for at least one value of the observed contents.

Win-Win

It is obviously impossible for both the casino and the player to be favored in the casino coin-tossing game. There must be an error in the reasoning of the casino, the player, or both. The game is fair only in the sense that it is symmetric. Matrix 5.1 gives what the player wins (and casino loses) for the specified number of tosses for the player and the casino.

The probability of the player tossing m times and the casino tossing n times is $(1/2^m)(1/2^n) = 2^{-m-n}$.

To each payoff in Matrix 5.1 we can associate its probability of occurrence $(1/2^m)(1/2^n) = 2^{-m-n}$, and we can replace each payoff with its weighted payoff, the product of its probability of occurrence and the payoff itself. For example, when the player tosses four times and the casino tosses three times the payoff is 8^3. This occurs with probability 2^{-7} giving a weighted payoff of $(2^{-7})(8^3) = 2^2 = 4$. The weighted payoffs are given in Matrix 5.2.

Again, we see the game is fair only in the sense that this matrix is symmetric. The sum of the payoffs in each column is positive and the sum of the payoffs in

Casino tosses n times

Player tosses m times

	1	2	3	4	5	→
1	0	-8^1	0	0	0	→
2	8^1	0	-8^2	0	0	→
3	0	8^2	0	-8^3	0	→
4	0	0	8^3	0	-8^4	→
5	0	0	0	8^4	0	→
↓	↓	↓	↓	↓	↓	

Matrix 5.1. Win-win payoffs.

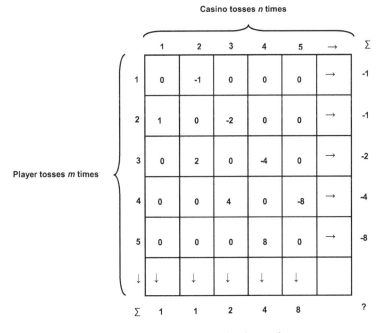

Casino tosses n times

Player tosses m times

	1	2	3	4	5	→	Σ
1	0	-1	0	0	0	→	-1
2	1	0	-2	0	0	→	-1
3	0	2	0	-4	0	→	-2
4	0	0	4	0	-8	→	-4
5	0	0	0	8	0	→	-8
↓	↓	↓	↓	↓	↓		
Σ	1	1	2	4	8		?

Matrix 5.2. Win-win weighted payoffs.

each row is negative. To determine the overall expected value of the game, we must compute the grand total. Adding columns first would give us partial sums diverging to $+\infty$; adding rows first would give partial sums diverging to $-\infty$. Because the summing process is order dependent, we conclude there is no sum; hence, there is no defined expectation for either the player or the casino. The game favors neither the player nor the casino and long-term average outcomes are unpredictable.

Parrondo's Paradox:
You *Can* Win for Losing

Two ugly parents can have beautiful children.
—Doron Zeilberger

I nspired by the behavior of mechanical ratchets, Spanish physicist Juan Parrondo of the Universidad Complutense de Madrid has shown that under precise circumstances games with negative (losing) expectation can be combined to give a winning result. So, losers listen up! Tired of searching for the clouds' silver linings? It may be the clouds themselves that offer hope. Before looking at Parrondo's paradox, let's take a quick look at how ratchets work.

Ratchets 101

A simple mechanical ratchet consists of a disk having a saw-tooth circumference and a spring loaded stopper, or pawl, designed to allow rotation of the disk in one direction only. (See Figure 6.1.)

Most readers are familiar with the ratchet wrench, a handheld tool that can tighten (or loosen) a bolt by repeated back and forth motions of the wrench. The pawl allows the bolt to turn only in the desired direction.

If random rotational forces are applied to the disk, the pawl blocks rotation in one direction, allowing rotation in the other direction. In a sense, the device draws order from randomness and can be put to work performing a task, such as winding a self winding watch. Figure 6.2 shows two chambers: The left chamber contains a mechanical ratchet, which allows the wheel to rotate in only one direction. The right chamber contains a paddle wheel being struck randomly by surrounding gas molecules. If the gas in the right chamber is heated to a sufficiently

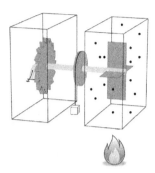

Figure 6.1. Ratchet and pawl.   **Figure 6.2.** Brownian motor.

high temperature, the paddles will rotate and be able to perform work such as lifting a small weight. This is a *Brownian ratchet* or *Brownian motor*, extracting a one directional force from the Brownian (random) motion of the gas molecules in the right chamber.

Parrondo's paradox is based on a third type of ratchet—a *flashing ratchet*, so named because continual switching, or flashing, is required between two mechanisms. Physical models of such mechanisms were used in the late nineteenth century as a kind of elevator, to lower and raise miners in and out of the tin mines of the Cornwall region of southwest England.

The Man Engines of the Cornwall Mines

On October 20, 1919, 31 tin miners were killed and many more injured when the Levant Mine of Cornwall, England, collapsed. The cause of the accident was the collapse of a so called *man engine*, a novel device used to efficiently lower miners into the mine and extract them when their shift came to an end. The device consisted of a continuous rod to which small step platforms were attached. It extended vertically into the mine, a distance of almost 2000 feet. Earlier models were driven by paddle wheel; the Levant engine operated by steam. The rod with attached steps would move up and down adjacent to a vertical column of fixed platforms. To descend into the mine, a miner would step from a fixed platform onto a step of the moving rod as it started its downward motion. When the downward motion stopped, the miner would step back onto a stationary platform and wait for the next downward cycle to occur. By systematically stepping back and forth from the stationary platforms to the moving platforms, miners could descend to the bottom of the mineshaft. Similarly, by riding the moving rod only on its upward cycles, miners could ascend. The two sets of platforms acted as a flashing ratchet, moving miners in the direction of choice, up or down, despite the fact that over time neither set of platforms achieved a net gain or loss in elevation. See Figure 6.3.

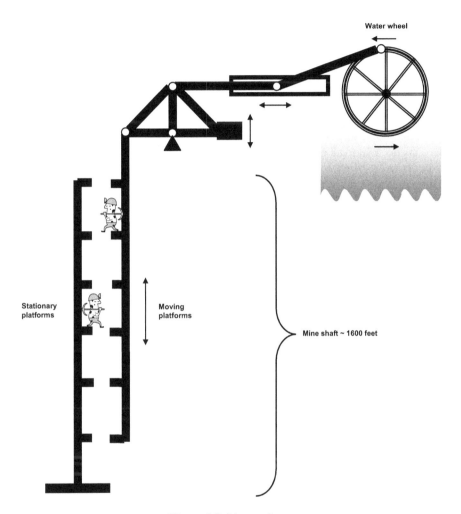

Figure 6.3. Man engine.

The catastrophe contributed to the mine's closing ten years later. The Levant engine, restored by the Cornish Engine Preservation Society and currently under the care of the British National Trust, is available for public view.

Such ratchet mechanisms inspired Spanish physicist Juan Parrondo to investigate the possibility of other phenomena behaving similarly. Parrondo discovered that under precise circumstances, systematically or randomly switching play between two losing games can result in an overall winning game. That is, there are games with negative expectations that can be played in combination so as to yield a positive expectation. Similarly, games with positive expectations can be combined to yield a loss. The general idea can be drawn from the preceding figure by imagining miners stepping back and forth so as to ascend from the mine as both sets of platforms sink very slowly downward, perhaps caused by a temporary settling of the ground below. The miners would be able to ascend, despite the overall downward motion of both platform columns. This would require the temporary settling to be sufficiently slow, as well as perfectly timed systematic switching by the miners. Parrondo gives examples where *random* switching between two losing games gives winning results.

Parrondo's discovery serves as a counterexample to the old adage of two wrongs not making a right. So, there's hope for losers! First we'll look at Parrondo's discovery, then consider similar phenomena, mathematical and otherwise. How can this be right when it feels so wrong?

Parrondo's Paradox

Gregory Harmer and Derek Abbott of the University of Adelaide, Australia, model Parrondo's discovery as a coin-tossing game consisting of two sub-games, game 1 and game 2, as defined below and in Figure 6.4 [Harmer and Abbott 99]. When played separately, each game is a losing game for the player. The expectations of both games are negative. Surprisingly, they can be combined to give a winning result.

Game 1

The player tosses a biased coin, with a probability of .49 of coming up heads and a probability of .51 of coming up tails. If it's heads the player wins $1, in which case the player's capital is increased by $1. If it's tails the player loses $1, decreasing the player's capital by $1. The expected value of game 1 is given by

$$E = .49(\$1) + .51(-\$1) = -\$.02.$$

Repeated plays of game 1 will result in an average loss to the player of 2 cents per game.

Game 2

Game 2 is composed of two sub-games, call them 2A and 2B. Game 2A is to be played if the player's capital is a multiple of 3. If not, game 2B is played. Each of the two games, 2A and 2B, involves the toss of a biased coin.

Game 2A

The coin is biased so as to come up heads with probability .09 and tails with probability .91. If it comes up heads, the player wins $1. If it comes up tails the player loses $1.

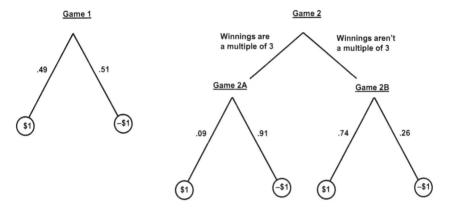

Figure 6.4. Two losing games.

Game 2B

The coin is biased so as to come up heads with probability .74 and tails with probability .26. If it comes up heads, the player wins $1. If it comes up tails the player loses $1.

So, whenever game 2 is played, it will be played as either game 2A or game 2B, most often as game 2B. A note of caution is in order when calculating the expectation of game 2 played by itself. Because game 2A is played whenever the winnings are a multiple of three, one might conclude game 2A accounts for one third of all plays of game 2. If we incorrectly assume the probabilities for game 2A and game 2B are respectively 1/3 and 2/3, then the expectation for game 2 appears to be positive,

$$E = \frac{1}{3}\big[.09(\$1) + .91(-\$1)\big] + \frac{2}{3}\big[.74(\$1) + .26(-\$1)\big] \approx \$.05,$$

making game 2 appear to be a winning game. In actuality, game 2A is played slightly more than 38 percent of the time, giving game 2 a negative expectation when played alone. Why is game 2A played more than 1/3 of the time?

Game 2A is played whenever the winnings are a multiple of three; the coin will almost always (91 percent of the time) come up tails and the player will lose $1. Now the winnings are no longer a multiple of three, and game 2B will be played with the player probably winning $1 and returning to a multiple of 3. Next, game 2A is played and it remains difficult to rise above the multiple of three. The "down then up" cycle may continue for some time, resulting in game 2A being played slightly more often than the originally estimated 1/3 of the time. Consequently, repeated plays of game 2 result in game 2A being played just over 38 percent of the time, giving game 2 an overall expectation of approximately –$.02. The game 2 player loses, on average, 2 cents per game. Details are given in this chapter's appendix.

Summarizing, game 1 has an expectation of exactly –$.02 per play and game 2 has roughly the same. Each game, when played repeatedly by itself results in an eventual loss to the player. These expectations are easily confirmed experimentally, by computer simulation. A discussion of game 2's negative expectation is given in this chapter's appendix.

Now consider playing game 1 twice in succession, followed by game 2 twice in succession, followed by game 1 played twice, and so on, repeating the pattern 1122. The unexpected expectation is positive! The precise calculation is complicated; but, simulation of the game shows the player should expect to win, on average, $.01 or 1 cent per game. Just as amazing is the fact that if the choice between the two games were to be made randomly, perhaps by the toss of a fair coin, the player's expectation is once again positive, approximately 1 cent per game. Best results are obtained by systematically switching periodically according to the pattern 12212, which yields an amazing expectation of $.06 or 6 cents per game. The most convincing argument is that of actually playing the games via simulation. Using the preceding diagram, simple programs are easily written to simulate game 1, game 2, and the various combinations of the two. The expectations are easily verified.

In some ways, the benefit of systematically switching should not surprise us. During a down cycle of the stock market, carefully timed switching between two losing stocks can generate profits if buys are made just prior to upswings and sells occur just before the drops. Buy low; sell high. It's easier said than done. Parrondo's paradox is surprising because the switching need not be systematic. In the long run you win, even without knowing precisely when to switch.

Reliabilism

In a paper entitled, "Parrondo's Paradox and Epistemology—When Bad Things Happen to Good Cognizers (and Conversely)," Fredrik Stjernberg of Linköping University, Sweden, gives epistemological applications of Parrondo's paradox [Stjernberg 07]. Stjernberg envisions levels of knowledge as steps on a ladder (see Figure 6.5), with an associated numerical scale, or metric. Zero corresponds to knowing nothing; the top rung describes a state of omniscience. Negative levels correspond to misinformation (falsehoods).

Process reliabilism justifies a belief by considering the process or steps used to arrive at the belief. Some processes are assumed to be more reliable than others. For example, mathematical proof, memory of recent events, and sensory perceptions are relatively reliable when compared to gross generalizations and hunches. Beliefs resulting from wishful thinking or paranoia are often false. Any reliable process generally takes one up the ladder; an unreliable process could allow a descent. Is a

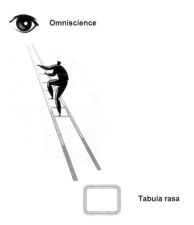

Figure 6.5. Levels of knowledge.

reliable process necessary to justify a belief? Is it sufficient? There is no consensus among philosophers. Stjernberg suggests the possibility of unreliable methods being randomly combined in a way that good results are obtained and the ladder is ascended, much like combining the two losing Parrondo games to create a winning game. He warns of the opposite, that two sound strategies may be naively employed in such a way as to slide down the ladder and away from the truth. We could dismiss this as nothing more than good (or bad) luck; but Stjernberg points out such results would be predictable, as with Parrondo's game.

Do the ends justify the means? Does knowledge, arrived at by unreliable methods, deserve to be called knowledge? One could argue that it does because the processes, combined in a Parrondo-like way, brought us up the ladder of knowledge, to a higher level of truth. What difference would it make that components of the complete process are unreliable?

From Soup to Nuts

Physical models of Parrondo's phenomena range, quite literally, from soup to nuts. Over three billion years ago the universe consisted of a random mixture of chemicals, a *primordial soup* (Figure 6.6), which yielded the complex amino acids necessary for protein molecules and life itself. Abbott suggests that Parrondo mechanisms may explain the extraction of order and life from primordial chaos [Abbott 09]. Randomness may have contributed to the process in a Parrondo-like way, much like random switching between game 2A and game 2B produces a profit from a loss. More than just being desirable, randomness may have been essential in creating the most wonderful unexpected expectation—life itself.

The probability of *abiogenesis* occurring by chance alone is quite small. British astronomer Sir Fred Hoyle compares it to the likelihood that a whirlwind

Figure 6.6. Add ocean water and stir.

blowing through a junkyard could produce a functional Boeing 747 [Hoyle 83, p. 19]. This comment by Hoyle arouses some controversy, a discussion of which is given in this chapter's appendix.

Too much speculation? Have scientists gone completely nuts searching for physical manifestations of Parrondo's paradox? Yes—Brazil nuts!

Parrondo's ratchet mechanism can be seen in something as simple as a bag of mixed nuts. Mix them up and the large Brazil nuts tend to rise to the top. They may be the largest and the heaviest of the collection, but rise, rather than fall, as a result of smaller nuts settling to the bottom and blocking the Brazil nuts from sinking. Breakfast cereals exhibit the same phenomenon—the *muesli effect*—if the box contains particles of different sizes. A biological example can be seen within cell walls, where enzymes and proteins position themselves in a beneficial way. Additional examples of the *Brazil nut effect* may be found in the fields of astronomy, geology, and geophysics, where vibrations mix particles triggering the ratchet effect.

Parrondo Profits

Can investment strategies be so modeled as to yield gains from investments that, by themselves, show losses? Mathematician and author John Paulos considers the possibility, referring to the potential gains as "Parrondo profits" [Paulos 03, pp. 52–54].

David Luenberger gives examples of *volatility pumping*, a ratchet-like technique for extracting profits from stocks showing no expected long-term growth [Luenberger 97, pp. 422–423]. A simple example will clarify things. We invest a sum of money, say $1, in a highly volatile stock whose value after one month is as likely to double as it is to be halved. If we buy and hold the stock, it rises and falls from month to month and there is no predictable long-term growth.

But, consider what happens when we invest $1 in the stock with the intention of cashing out after one month. There is a 50 percent chance of doubling our money and a 50 percent chance of losing half of it. So, the expected value of the stock when we sell would be positive:

$$E = \frac{1}{2}(\$2.00) + \frac{1}{2}(\$.50) = \$1.25.$$

This can be exploited by periodically removing a fraction of the capital invested in the stock and saving it as cash. In this case, assume we rebalance at the beginning of each month so that half of our capital remains invested in the volatile stock and half is saved as cash. In a good month our volatile stock will double and our total capital will increase by a factor of 3/2. In a bad month the volatile stock will lose half its value and our total capital will lose one fourth of its value. So, after t months, we should expect our investment to increase by a factor of 3/2 approximately $t/2$ times and lose one fourth of its value an equal number of times. A reasonable expectation for the value of our $1 investment after t months would be $(3/2)^{t/2}(3/4)^{t/2} = (3/\sqrt{8})^t$, growing exponentially at a rate of approximately 6 percent per month. A similar analysis shows that monthly rebalancing two equally volatile investments, each having a 50 percent probability of doubling or halving after each month, yields a long-term unexpected growth rate of approximately 11.8 percent per month! Most amazing is the predictable long-term growth achievable from two losing stocks, assuming both are sufficiently volatile. If in the above example we assume that each of our two stocks doubles with probability 49 percent and halves with probability 51 percent, then it is easily shown that rebalancing as above yields long-term exponential growth at a rate of approximately 10 percent per month, despite the fact that both stocks are individually long-term losers.

Volatility pumping derives growth from volatility in much the same way as Parrondo games produce wins from losing games and ratchets extract one directional motion from random motion. It is indeed true that ugly parents can have beautiful children.

Truels: Survival of the Weakest

A *truel* is the three-person equivalent of a duel. Picture three players positioned at the vertices of an equilateral triangle. In a sequential truel each player shoots at one of two opponents in a predetermined order, with the objective of surviving and eliminating as many opponents as possible. The ability (marksmanship) of each player is given by a probability measuring the likelihood of a single shot hitting and killing its target. Assuming the players each have unlimited ammunition, play continues until there is a single survivor. In some formulations the players are each given a finite number of bullets and play stops when either a single survivor or no bullets remain. Some versions allow for a player to intentionally miss (a misfire or abstention) by firing into the ground. In all versions the shooter must decide at which opponent to aim, or to intentionally miss. If utilities are assigned

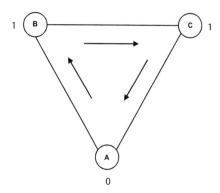

Figure 6.7. Sequential truel.

to survival and the elimination of opponents, the truel can be discussed in terms of mathematical expectation. Our brief analysis will simply consider survival as the primary objective and the elimination of opponents as secondary.

Truels are not Parrondo games; however, they clearly illustrate the benefits of a "losing to win" strategy. There can be strength in weakness and in some cases the weakest player has the best chance of survival. Unexpected optimal strategies yield unexpected outcomes.

As a first example, assume A is the weakest of the three, never hitting the target; A's marksmanship is therefore 0. Players B and C each have perfect marksmanship, hitting their chosen targets with certainty. All are given one bullet each and assume they agree on the firing order A → B → C. Figure 6.7 shows the players with their respective marksmanships. The firing order is indicated by arrows, and we assume A fires first.

Is A doomed? To the contrary, A, by far the weakest player, has a better chance of survival than B or C. In fact, A is sure to survive! To see why, let's walk through the play. A fires first and misses; A's target is irrelevant. It is then B's turn to fire and C should be the chosen target because A is not a threat. B kills C and play ends with A and B surviving.

If the selected firing order is C → B → A, then once again A will survive. Playing it out, C kills B, A fires, misses, and play ends. The survivors are A and C.

Similar reasoning shows that no matter which of the six possible firing orders is agreed upon, A will always survive (see Table 6.1). Neither B nor C will consider A to be a threat; instead, they will shoot at each other, allowing for A's survival.

If the firing sequence is randomly selected from among the six possible sequences, A survives with probability 100 percent, B and C each survive with probability 50 percent. Though significantly the weakest player, A is twice as likely to survive as each of its two opponents. It's literally *survival of the weakest!*

The above analysis hints that, given the right circumstances, a player might be best off wasting the first shot by firing into the ground, effectively becoming

Sequence	Surviving Players
A→B→C	A and B
A→C→B	A and C
B→A→C	A and B
B→C→A	A and B
C→A→B	A and C
C→B→A	A and C

Table 6.1. Survival of the weakest.

the worst player. To see why, consider that A, B, and C all have perfect marksmanship; they never miss. Each is given one bullet and they agree on a firing order, say A → B → C. Firing first, A would be wise to fire into the ground. Suicidal? Again, and to the contrary, this guarantees A's survival. Had A chosen to kill B or C, then the remaining player would surely have killed A, with play ending then. By firing into the ground, B must kill C, else be killed by C, and play ends. So, for any predetermined firing order, it would be wise for the first player to waste the shot.

Kilgour and Brams discuss sequential truels for which every shooter is randomly selected from the survivors, allowing the possibility of a player firing multiple times in succession [Kilgour and Brams 97]. Assume there is an unlimited supply of ammunition and intentional misses are not allowed. (This rules out the possibility of a three way mutual abstention where all three players survive.) Also assume that when all three players are present, the shooter aims at the stronger of its two opponents, so as to minimize the risk of being killed by an opponent. Play continues until a single player survives. Six sets of marksmanships are given in Table 6.2 with the survival probabilities of each player. For all six, A is the least likely to hit the target and C is the most likely to do so.

All six orders of the survival probabilities are possible, depending on the marksmanships. In the first set, the survival probabilities are in the order of the

Marksmanship			Survival Probabilities			Order of Survival Probabilities
A	B	C	p_A	p_B	p_C	
0.1	0.5	0.9	0.127	0.333	0.540	$p_A < p_B < p_C$
0.3	0.5	0.9	0.309	0.294	0.397	$p_B < p_A < p_C$
0.2	0.7	0.8	0.212	0.412	0.376	$p_A < p_C < p_B$
0.25	0.5	0.6	0.316	0.370	0.314	$p_C < p_A < p_B$
0.4	0.5	0.8	0.392	0.294	0.314	$p_B < p_C < p_A$
0.4	0.6	0.8	0.370	0.333	0.296	$p_C < p_B < p_A$

Table 6.2. Survival probabilities.

players' marksmanships. In the last and most surprising case, A, the weakest, is most likely to survive and C, the strongest, is least likely. Survival of the weakest! The computation of the probabilities for this last case is given in the appendix.

It's natural to envision a truel as a Wild West shootout or some other form of dramatic entertainment. Portrayals appear in such popular films as *The Good, the Bad and the Ugly, Pirates of the Caribbean: Dead Man's Chest, Reservoir Dogs,* and *Pulp Fiction.* But real-world models exist as well. Abbott models the interaction of political parties using truels, noting that large governments are often dominated by a two party system. Third parties may benefit by staying out of the fray, allowing the two major parties to fight it out amongst themselves as a duel. In any case, entering the fray could prove costly and may hurt the major party with which a third party may be most aligned. The example cited is that of Republican candidate George Bush beating Democratic candidate Al Gore in the US presidential election of 2000. Green Party candidate Ralph Nader finished with 2.5 percent of the vote. Some suggest that Nader would have been better off "firing into the ground" and dropping out early. Had he done so, the vast majority of his votes would have gone to Gore (more aligned with Nader's political philosophy than Bush), possibly changing the election's outcome. By staying in to the bitter end, Nader may have hurt his cause and that of the Green Party.

Abbott speculates similarly that marriages consisting of two partners tend to survive longer than those consisting of three or more partners. And why are there only two sexes for sexual reproduction and not three (or more)? It's something to think about.

Going North? Head South!

Parrondo games, volatility pumping, reliabilism, and truels all derive forward motion from random or reverse motion. It's like watching a Michael Jackson moonwalk in reverse! Sometimes what feels wrong may yet be right. A most striking example of this is taken from the *calculus of variations* and serves as the concluding topic in this chapter. Neither mathematical expectation nor probability is involved in the discussion to follow. But the optimal solution is highly unexpected and is reminiscent of discussions given in this chapter. Sometimes the quickest way up is down!

The problem is that of choosing a continuous path between two points, A and B, which will minimize the time of travel between the two points. The straight line segment joining the two points may, in the Euclidean sense, minimize the distance; but, for various reasons, this may not always serve as the path of minimal time. There may be position and speed constraints that must be satisfied. When you walk from your desk at home to your refrigerator in the kitchen, do you travel in a straight line? Unlikely! Walls prevent a straight-line path, unless your intention is to bash your way through them on the way to the kitchen. In the unlikely event you grab a sledge hammer and proceed in this way, travel distance

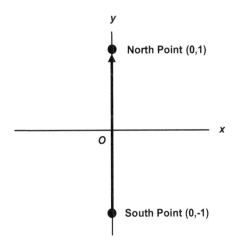

Figure 6.8. South Point to North Point, minimizing distance.

will be minimized but travel time will certainly not be minimized. Wouldn't it have been quicker (and less disruptive to family members) to simply walk into the kitchen in the usual way? When traveling from Los Angeles to London, a straight line segment path is completely out of the question, unless you plan on digging through thousands of miles of the earth's crust to reach your destination. You reach your destination in minimal time by traveling (probably flying) along the path of a great circle on a sphere. Weather and air traffic considerations may impose additional constraints, requiring the course to be further altered, adding miles to your journey so as to make the trip in minimal time.

Formally, problems of this type are solved using calculus-of-variations techniques. Presented here is a simple problem with a highly unexpected solution, again reminding us that what feels wrong may be right.

You are positioned at South Point, located at $(0,-1)$ in the Cartesian (xy) plane. You must choose a continuous path in the plane so as to arrive at North Point, located at $(0,1)$, in minimal time. With no position or speed constraints, minimal distance and minimal time would be achieved by heading straight north, through the origin $(0,0)$, ultimately arriving at North Point as shown in Figure 6.8.

But, let's complicate matters by imposing a speed constraint. Travel is treacherous in the neighborhood surrounding the origin (perhaps due to bad road or traffic conditions) and speed in the plane is determined by the radial distance from the origin. Specifically, speed = $x^2 + y^2$, the square of the distance from the origin. So, starting at South Point, your initial speed is 1, and you would slow down if you were to head straight north, closer to the origin. At $(0,-1/2)$ your speed would only be $0^2 + (-1/2)^2 = 1/4$. You are free to turn in any direction, but you must move along a continuous path and your speed is strictly determined by your position.

Which course would you take so as to arrive at North Point in the least amount of time? The answer is remarkably counterintuitive.

First realize that heading straight north is futile as you would never even reach (0,0), let alone (0,1) in finite time. To see why reaching (0,0) is impossible, consider attempting to do so and note that the time required to reach (0,−1/2) would exceed 1/2 because the greatest speed over that interval would be 1. The time required to travel from (0,−1/2) to (0,−1/4) would be greater than 1 as the greatest speed over that interval would be 1/4. Traveling from (0,−1/4) to (0,−1/8) would be at least 2 because the maximum speed over that leg would be 1/16. Halving the distance to (0,0) at each step, infinitely many segments would be traveled requiring a total time of more than 1/2 + 1 + 2 + 4 + ... , a sum that is infinitely large. The journey could not be completed in finite time. A more technical proof is given in this chapter's appendix.

To minimize the time from South Point to North Point, you could move away from the origin to gain speed and then circle around to the north, finally dropping back down to North Point (see Figure 6.9). To be precise, you could initially head south to some point, say (0,−r), r > 1, then travel around on a semicircle to the point (0,r). Finally, head straight south for the final leg of the journey to North

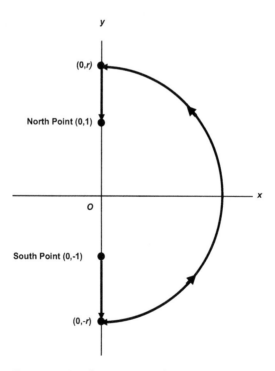

Figure 6.9. South Point to North Point in minimal time.

Point. The time required for this journey is $2 + \frac{\pi - 2}{r} > 2$, which can be made arbitrarily close to 2 as r approaches infinity. Details are given in the appendix.

To summarize, the best you can do to minimize the duration of the journey is to initially head south, *in the opposite direction of your destination*. How far south should you head before swinging around to the north? The further the better. And, the further south you go the faster you will be initially traveling in the exact opposite direction of your destination, North Point. The magnitude of the sacrifice is difficult to comprehend.

Positioned at South Point you ask a local resident for directions to North Point, two miles due north:

YOU: Which is the quickest route to North Point?
LOCAL: Head south for just under one hour, then turn left and circle back up to the north. You can make the trip in just over two hours.
YOU: Head south to North Point?
LOCAL: Yeppers!
YOU: How far south?
LOCAL: Very far.
YOU: You're not making sense.
LOCAL: Sorry. You asked for directions and I'm giving them to you.
YOU: How will I know when to turn left?
LOCAL: You won't. Watch the time and just before it reaches one hour make the turn.
YOU: What if I wait too long?
LOCAL: Ooh, not good!
YOU: Can I make it on one tank of gas?
LOCAL: What kind of mileage do you get?
YOU: I burn about 2 gallons of gas per hour at high speed.
LOCAL: You'll make it easily on one tank.
YOU: I think I get about 25 miles per gallon.
LOCAL: In that case, forget about it!

Appendix

The Expectation of Game 2 (Parrondo)

Consider game 2 as a three-state *Markov chain*, a mathematical system that transitions from one state to another. State 1 represents the player's accumulated capital one unit above a multiple of three. State 2 indicates an accumulated capital of two units above a multiple of three. Finally, state 3 denotes an accumulated capital that is a multiple of three. So, for example, the probability of moving from state 2 to state 1 in one play of game 2 is .26. The probability of remaining in state 2 after one play is 0 because one play must result in a win or a loss. The transition matrix, which displays the probabilities of moving from one state to another, is formed from the probabilities given earlier in this chapter:

$$
T = \begin{array}{c@{\quad}c@{\quad}c@{\quad}c}
 & 1 & 2 & 3 \\
1 & 0 & .74 & .26 \\
2 & .26 & 0 & .74 \\
3 & .09 & .91 & 0
\end{array}.
$$

Note that the columns and rows are labeled with the possible states. For any given initial condition P_0 (in the form of a vector) and for n plays of the game, $\lim_{n \to \infty} P_0 T^n \approx (.155 \quad .463 \quad .383)$, showing that the probability of being in state 3 stabilizes at approximately 38.3 percent. (The result of such a multiplication is known as the *state distribution vector*.) Under these conditions, the expected value associated with playing game 2 by itself is approximately –$.02, a losing game. (A relatively elementary discussion of Markov chains can be found in almost any finite mathematics textbook, e.g., [Gilligan and Nenno 79, pp. 265–278].)

Hoyle's Junkyard Tornado

The distinguished British astronomer and author Sir Fred Hoyle (1915–2001) is known for significant contributions to astrophysics, specifically stellar evolution. He will also be remembered for his controversial opinions regarding cosmology and the origin of life. He coined the expression "Big Bang" to mock physicist George Gamow's theory that the universe began ten to twenty billion years ago as an explosion of intense heat, creating matter and the forces of nature in less than one second. Hoyle proposed an alternative *steady state theory*. According to Hoyle (Fred, not Edmond), the universe has always existed, continuously expanding. The Big Bang theory is generally accepted today, yet remains known by Hoyle's pejorative description.

Hoyle opposed the idea of life originating by chance from a primordial soup, believing that such an event was so unlikely that its likelihood was close to that of a tornado blowing through a junkyard and creating a functional Boeing 747. In place of abiogenesis, Hoyle promoted *panspermia*, the theory that life has always existed in the universe and came to the Earth from elsewhere, traveling through the vacuum of space. Hoyle's theory is open to criticism. In any event, it dodges the issue of life's origin by simply suggesting life came from somewhere else.

Some creationists and believers in intelligent design adopt Hoyle's 747 analogy as evidence of God's creation of life, their argument being that "God's work" is a far more plausible and likely explanation of life's origin than abiogenesis. Modern evolutionary biologists generally reject creationist views and Hoyle's 747 argument. Famed biologist, author, and avowed atheist Richard Dawkins describes

Hoyle's argument as "Hoyle's Fallacy." Dawkins writes, "However statistically improbable the entity you seek to explain by invoking a designer, the designer
himself has got to be at least as improbable. God is the ultimate Boeing 747"
[Dawkins 06, p. 138].

Random Truel Survival Probabilities

The survival probabilities for the last case of the random truel are calculated by considering the truel to be a seven-state Markov chain with states ABC, AB, BC, AC, A,
B, and C reflecting the survivors after n shots have been fired. The initial state is ABC,
and the chain terminates in one of the three absorbing states A, B, or C.

The transition matrix is

$$T = \begin{array}{c c} & \begin{array}{ccccccc} ABC & AB & BC & AC & A & B & C \end{array} \\ \begin{array}{c} ABC \\ AB \\ BC \\ AC \\ A \\ B \\ C \end{array} & \left(\begin{array}{ccccccc} .4 & .\overline{3} & 0 & .2\overline{6} & 0 & 0 & 0 \\ 0 & .5 & 0 & 0 & .2 & .3 & 0 \\ 0 & 0 & .3 & 0 & 0 & .3 & .4 \\ 0 & 0 & 0 & .4 & .2 & 0 & .4 \\ 0 & 0 & 0 & 0 & 1 & 0 & 0 \\ 0 & 0 & 0 & 0 & 0 & 1 & 0 \\ 0 & 0 & 0 & 0 & 0 & 0 & 1 \end{array} \right) \end{array}.$$

The initial state is represented by $\begin{pmatrix} 1 & 0 & 0 & 0 & 0 & 0 & 0 \end{pmatrix}$. As the number of shots
fired increases, the state distribution vector approaches

$$\lim_{n \to \infty}\begin{pmatrix} 1 & 0 & 0 & 0 & 0 & 0 & 0 \end{pmatrix} T^n \approx \begin{pmatrix} 0 & 0 & 0 & 0 & .370 & .333 & .296 \end{pmatrix},$$

giving the survival probabilities in the reverse order of the players' marksmanships.

South Point to North Point in Minimal Time

This explanation requires calculus, which usually comes into play when dealing
with minimums or maximums. In our example, the speed is defined by the (x,y)
position as $x^2 + y^2$. The point $(0,0)$ can't be reached in finite time traveling straight
up from $(0,-1)$. With $x = 0$, $y^2 =$ (change in distance)/(change in time) $= dy/dt$,
and the time required is given by $t = \int_{-1}^{0} \frac{1}{y^2}dy$, an integral that diverges to infinity.

Joel Lewis offers a clever, sidewise-thinking approach to reaching North Point in minimal time. Think of $P = (x,y)$ as a point moving in the Cartesian plane. To each point P associate an image point

$$Q(x,y) = \left(\frac{x}{x^2 + y^2}, \frac{y}{x^2 + y^2} \right).$$

The speed of point P is given by $x^2 + y^2$. That is $\sqrt{(dx/dt)^2 + (dy/dt)^2} = x^2 + y^2$. (The left-hand side is a standard calculus representation of speed. For example, if the x coordinate increases at 3 inches per second ($dx/dt = 3$) and the y coordinate increases at 4 inches per second ($dy/dt = 4$), then the speed of the particle across the xy plane is 5 inches per second.) The equality of the two expressions is used in the derivation to follow.) We now show that the speed of the image point Q is always 1, except at (0,0) where the speed of Q is undefined.

The speed of Q is

$$\sqrt{\left[\frac{d}{dt}\left(\frac{x}{x^2+y^2} \right) \right]^2 + \left[\frac{d}{dt}\left(\frac{y}{x^2+y^2} \right) \right]^2}$$

$$= \sqrt{\frac{\left[\frac{dx}{dt}\left(x^2+y^2 \right) - x\left(2x\frac{dx}{dt} + 2y\frac{dy}{dt} \right) \right]^2}{\left(x^2+y^2 \right)^4} + \frac{\left[\frac{dy}{dt}\left(x^2+y^2 \right) - y\left(2x\frac{dx}{dt} + 2y\frac{dy}{dt} \right) \right]^2}{\left(x^2+y^2 \right)^4}}$$

$$= \sqrt{\frac{\left(x^2+y^2 \right)^2 \left[\left(\frac{dx}{dt} \right)^2 + \left(\frac{dy}{dt} \right)^2 \right]}{\left(x^2+y^2 \right)^4}} = \sqrt{\frac{\left(x^2+y^2 \right)^4}{\left(x^2+y^2 \right)^4}} = 1,$$

assuming $(x,y) \neq (0,0)$.

As P travels on a continuous path from (0,−1) to (0,1), so must Q, doing so with unit speed. The time of Q's journey is minimized by Q traveling the straightest possible path to (0,1) avoiding (0,0) by an arbitrarily small distance. In doing so, the time of travel will always exceed 2; but the difference can be made arbitrarily small. One option corresponds to P initially heading down from (0,−1), arbitrarily far to a point (0,−r), $r > 1$, then circling around to (0,r), finally dropping down to the target (0,1).

Imperfect Recall

Your memory is a monster; you forget—it doesn't. It simply files things away. It keeps things for you, or hides things from you and summons them to your recall with a will of its own. You think you have a memory; but it has you!

—John Irving

You've just spent two hours shopping at your local mall and are now heading out into the parking lot, searching for your car. After walking up and down two lengthy aisles of parked cars with no success, you must now decide whether to keep walking or return to the mall and ask security for assistance. Continuing to search is a waste of time if, in fact, your car has been stolen. On the other hand, you may have simply forgotten its location and will suffer some embarrassment if mall security must locate your car for you. What do you do?

Imperfect recall, as it applies to solitaire (single-player) games, is the inability of the player (perhaps due to absentmindedness) to recall past actions. This may result in the player having *imperfect information,* the inability to distinguish between nodes (decision or branch points) on the game's extensive (tree) form. Your dog sits patiently at his food bowl and you can't recall, with certainty, whether or not he has been fed. Do you feed him now, running the risk of overfeeding and upsetting his stomach, or ignore the little fellow, knowing that he may have missed a meal and will go hungry until the following day?

Or, consider the problem of taking a prescribed medication four times a day and you can't recall how many pills you've already taken. What should you do? You can take a pill now, running the risk of overdosing with possible side effects, or skip the dose with the risk of under dosing and losing the medication's benefit. (Some antibiotics are packaged so each dose is individually marked in the packet, eliminating the problem of imperfect recall.)

We could analyze the above problems by assigning probabilities and utilities to the decision trees and make our decision so as to maximize the expected value of our choice. A hypothetical scenario by Piccione and Rubenstein raises some interesting questions and paradoxes [Piccione and Rubinstein 97]. Since the seminal paper's publication, extensive discussion of imperfect recall appears in academic journals and on the Internet. This chapter begins with Piccione and Rubinstein's "The Absentminded Driver," followed by a popularized Internet variation, "The Sleeping Beauty Problem."

The Absentminded Driver

After work on Friday, a man stops at a local pub for a few drinks. To get home safely (maximum payoff), he knows he must head south on the highway and take the second exit from his present location. If he incorrectly takes the first exit, he will be forced to drive through hazardous road conditions in order to return to the highway and head home. If he misses both the first and the second (correct)

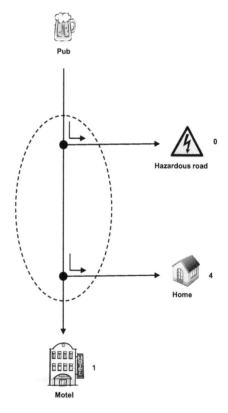

Figure 7.1. The absentminded driver.

exit, he will reach a dead end miles down the road and be forced to spend the night at a motel before returning home the following morning. Payoffs proportional to the consequences are given in Figure 7.1. From his perspective, the best outcome is to arrive home by taking the second exit. To this we assign a payoff of 4. The next best outcome would be for him to miss both exits and be forced to spend the night at a motel. This is assigned a payoff of 1. The worst possible outcome is to take the first (hazardous) exit. Since this could result in bodily injury and damage to his car, a payoff of 0 is assigned.

To complicate matters, he anticipates being intoxicated on his drive home. Both exits will appear identical and he will be unable to distinguish between the two, not being able to recall if he bypassed an exit. The dashed oval in Figure 7.1 indicates his inability to distinguish between the two exits. He cleverly decides to plan his drive home now, while relatively sober, and then stick to it when he heads home. "Plan the drive and drive the plan!"

During this planning, or *ex ante*, stage, he reasons as follows. Whenever he arrives at an exit, he must decide whether or not to take it. Both exits will appear identical, and he will not know at which exit he is positioned. He may as well decide now whether to exit or go straight. His decision applies to both intersections, as they will be indistinguishable. If his decision is to exit, then he will exit at the first opportunity with dire consequences, corresponding to a payoff of 0. If he goes straight, he will do so at both opportunities, arriving at the motel for a payoff of 1. At the planning stage it's a no brainer. He should go straight and spend the night at the motel.

This now becomes the man's Friday routine. He stops off for a few drinks at the pub, drives the plan by going straight until the road ends, and spends the night at the motel, returning home on Saturday morning. A payoff of 1 is associated with each occurrence.

On one of these Friday evenings, while preparing to bypass the upcoming exit, he does a quick mental expected value calculation. The approaching exit is either the first or the second, each with a probability of 1/2. If, contrary to past practice, he were to exit now, the unexpected expectation would be

$$E = \frac{1}{2}(0) + \frac{1}{2}(4) = 2,$$

twice the payoff of going straight. How can this be? Why has he been going straight and spending Friday evening's away from home when he could double the expected payoff by taking the exit? Being too drunk to resolve the issue while driving, he continues going straight and once again spends the night at the motel. He vows to resolve the problem the following morning when he's sober.

There appears to be two contradictory optimal strategies, one at the planning stage and one at the action stage while driving. At the pub, during the planning stage, it appears the driver should never exit. But once this plan is in place and he arrives at an exit, a recalculation with no new significant information shows that exiting yields twice the expectation of going straight.

Next Friday our driver heads home from the pub. But now he's in a quandary as to going straight or exiting. If he decides to exit, then he will do so at the first possible opportunity with catastrophic results. If he decides to go straight, as he has done so many times before, then he should quickly revise his strategy and exit. But if he exits, ... And so it goes. What would you do if you were the absent-minded driver?

Two years ago I spoke to a group of California community college mathematics instructors at a conference in Monterey, California. Part of the talk was devoted to the absentminded driver problem. My wife, seated in the audience, overheard two instructors discussing the dilemma. One asked the other, "Why doesn't he just get a GPS?" Good point!

A mixed strategy formulated at the planning stage does not resolve the dilemma. Let's see what happens if we allow the driver to randomize his exit decision. If he exits with probability p, then

$$E = p \cdot 0 + p(1 - p) \cdot 4 + (1 - p)^2 \cdot 1 = -3p^2 + 2p + 1.$$

This is maximized at $p = 1/3$, giving an expectation of $E = 4/3$. This payoff is superior to that of his pure (planning-stage) strategy of going straight and avoids the paradox associated with revisions. He decides to use the randomizing strategy, exiting with probability 1/3 and keeps a random number generator in the glove compartment of his car. For the next few Fridays, all goes as expected and his average payoff is now 4/3.

Several weeks later, sitting in the pub waiting for his first beer to arrive, he grabs a cocktail napkin and makes a few quick calculations. For every three times he reaches the first exit, he will continue straight two times, reaching the second exit twice, meaning that five exits will be reached in total. It follows that whenever he arrives at an exit, the probability of it being the first exit is 3/5 and the probability of it being the second is 2/5. Under these conditions, what if he revises his strategy so as to exit without randomizing?

His expectation is

$$E = \frac{3}{5}(0) + \frac{2}{5}(4) = \frac{8}{5} > \frac{4}{3},$$

indicating that he is better off exiting with certainty as opposed to randomizing. But this is based on his planning stage strategy of randomization and the earlier paradox now reappears.

Barton Lipman of the University of Western Ontario comments on the absentminded driver problem [Lipman 97, pp. 97–100]:

> It is hard to believe that a game with one player and one information set could inspire so many eminent and intelligent people! ... I believe the fundamental contribution of Piccione-Rubinstein is to show that some very compelling intuitions which all lead to the same conclusion in games with perfect recall lead to different conclusions with imperfect recall. This surprising fact, which some label a "paradox," forces us to reconsider these different intuitions to find where the differences lie. Once we identify these differences precisely (as many people have done with admirable creativity and clarity), we end up constructing a series of internally consistent and different views of how to model an agent with imperfect recall—and gaining a new understanding of things we thought we already understood completely.

Robert Aumann, Sergiu Hart, and Motty Perry offer a practical solution [Aumann, Hart, and Perry 97]. The driver's dilemma is caused by lack of recall and his inability to distinguish at which exit he is currently positioned. To improve his expectation he devises the following plan. Whenever he bypasses an exit, he will toggle his car radio on (if currently off) or off (if currently on). He will leave the pub with a probability of 1/2 that the radio is on. (He could toss a coin and turn the radio on if it comes up heads.) As he heads down the highway, his lack of recall will not allow him to remember the initial state (on or off) of his radio. His plan is to exit if the radio is on and bypass an exit if the radio is off.

Toggling the radio on and off, assuming we allow this, does not violate the assumption of imperfect recall. Since the driver does not know the initial state of his car radio and has no recall of previous actions, he will not be able to distinguish which of the two exits he is approaching. The condition of imperfect recall remains satisfied. The toggling strategy forces the driver to take one of the two exits, each with probability 1/2, without the driver knowing which exit he is taking. His expectation using this strategy is

$$E = \frac{1}{2}(0) + \frac{1}{2}(4) = 2,$$

a payoff without paradox and no worse than that of any other strategy.

Better yet, as suggested by the aforementioned conference attendee, "Buy a GPS!"

Unexpected Lottery Payoffs

There are endless variations of the absentminded driver problem, each with its own curiosities. (One variation shows a player with more information may do worse than one with less [Board 03].) Framing such problems in terms of deciding

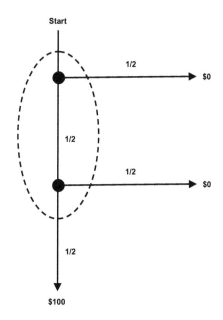

Figure 7.2. Unexpected lottery payoff, Example 1.

whether or not to purchase a lottery ticket yields unexpected results [Aumann, Hart, and Perry 97].

You are a passenger in a car beginning your journey as shown in Figure 7.2. You, as passenger, have no control whether or not to turn at each of the two intersections, which Aumann, Hart, and Perry refer to as *nodes*. The two nodes are shown as large, black dots. The probability of a turn occurring at any given node is 1/2, as shown on the figure. As usual, the dashed oval indicates your inability to distinguish between the two nodes. This could be caused by dense fog, your being unfamiliar with the territory, or severe intoxication. At the start, and at each of the two nodes, you will be given the opportunity to purchase a lottery ticket for $30, which will yield a payoff of $100 only if no turns are made and the journey ends at the bottom of the diagram. If a turn occurs at either node, the payoff is $0, as shown.

From the start, the $100 payoff occurs with probability $1/2 \cdot 1/2 = 1/4$ and the expected payoff is therefore $25. If, at the start, the ticket is purchased for $30, then the expected value of the purchase is $25 − $30 = −$5, an expected loss of $5. But what if you're offered the ticket at the first node? Positioned at the first node, you are unable to determine if you are at the first node or the second. But, since the first node receives twice as much traffic (when all players are considered) as the second, you conclude that your probability of being at the first node is 2/3 and your probability of being at the second is 1/3. Consequently, your expected payoff (ignoring the cost of the lottery ticket) is

$$E = \frac{2}{3}\left(\frac{1}{2}\right)^2 (\$100) + \frac{1}{3}\left(\frac{1}{2}\right)(\$100) = \$33\frac{1}{3}.$$

Purchasing the ticket for $30 at the first node seems favorable, associated with a gain of $3\frac{1}{3}$. This is odd. You reach the first node with certainty and your information does not significantly change there. Yet, your expectation changes from a losing bet to a winner. Are you better off purchasing the ticket at the first node than at the start? From your perspective the answer is "yes."

Think of it this way. The purchase is unfavorable if purchased at the start. If, in actuality, you're at the first node, then the purchase would be equally unfavorable there as well. But, because the first and second nodes are indistinguishable to you, you have no way of knowing whether you're at the first (unfavorable) node or the second (favorable) node. Because you believe there is a significant probability of being at the second node, you perceive (correctly) that, on balance, the purchase as favorable.

Figure 7.3 is another example of an unexpected lottery payoff. Once again you are the passenger in a car and have no control as to which of three roads is taken. The probability is 1/3 of going down any one of the three roads shown.

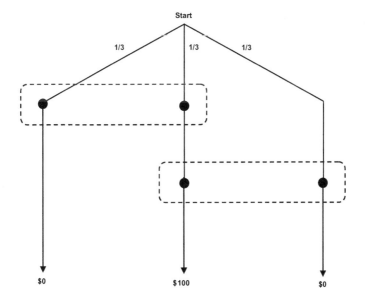

Figure 7.3. Unexpected lottery payoff, Example 2.

The large black dots, or nodes, may be considered as rest stops. As always, each of the dashed rounded rectangles, or *information sets*, contain two indistinguishable nodes. There are two such information sets. The imperfect recall condition assumes you will not be able to recall being in a previous information set, if this were to be the case. Note that this could only occur if the middle of the three roads is taken. The lottery will pay $100 only if the middle road is taken.

At the start, there is a 1/3 probability of winning $100 and your expected payoff is 33\frac{1}{3}$. But when you arrive at an information set you will perceive your probability of being on the middle road to be 1/2. So your probability of winning $100 is 1/2, yielding an expected payoff of $50. Would you be willing to pay $40 at the start for such a ticket? Would you pay $40 at one of the information sets for the ticket?

A real-world model of this lottery is conceivable. A six-sided fair die with sides labeled L, L, M, M, R, and R is rolled once to randomly choose one of the three (left, middle, or right) roads. The middle road pays you $100 at the end of the experiment. The left and right roads pay you nothing. You are not to know the outcome of the roll until the end of play. Whenever an information set is reached, you are told only that you are in that set. For example, there may come a time when you are advised, "You are in information set LM," indicating that you are on road L or road M, each with probability 1/2. Or, there may come a time when you are told, "You are in information set MR," indicating that you are on road M or road R, each with probability 1/2. Assume you are sufficiently forgetful (or drugged) so that if on road M and passing through information set MR, you will not recall having passed through information set LM.

Now consider this. At the start, you purchase the ticket for $40. (You should have no trouble finding someone willing to sell you a ticket at this price.) And why should you purchase such a ticket? *Because you know, with certainty, that you will correctly perceive the value of the ticket to be $50 once you arrive at an information set, which you will do with certainty. And, at that point, you should be able to sell the ticket at a profit, to anyone else privy to the same information.* Buy low, sell high? No problem!

Is there a similar, real-world, get-rich-quick scheme? Perhaps not. One might dismiss the possibility by dismissing games of imperfect recall and imperfect information as mathematical contrivances having little to do with reality. But, in actuality, games of perfect recall and information are the idealized contrivances. We're generally aware of an infinitesimally small fraction of existing information and recall only a small fraction of our past actions. So, as bizarre as the preceding lottery examples may appear, there is the possibility of real-world models.

Sleeping Beauty

The *sleeping beauty problem* first appeared in a paper by Adam Elga, published in *Analysis* [Elga 2000]. It can be appreciated by anyone with a minimal knowledge of probability and has been the topic of many heated Internet discussions.

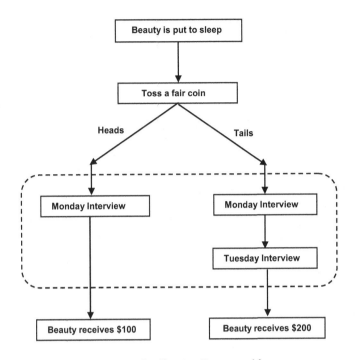

Figure 7.4. The Sleeping Beauty problem.

Sleeping Beauty is put into a drug induced sleep and will be awakened once (Monday) or twice (Monday and Tuesday) depending on the outcome of a coin toss (heads once, tails twice). After each waking, she will be put back to sleep by a drug causing her to forget that waking. At each waking Beauty will be asked, "What is your credence now that the coin came up heads?" (We take *credence* to be her confidence, or subjective estimate of the probability.) We assume Beauty fully understands the nature of the experiment and the notion of probability. She will not know whether it is Monday or Tuesday and will not be able to recall whether or not she has been previously interviewed. How should Beauty answer the question?

The usual formulation of the problem requires Beauty give a probability for an answer. If Beauty receives $100 per interview (one or two) for her efforts, the question can be posed in terms of expected value by asking Beauty, "How much money now do you expect to collect at the end of the experiment?" The experiment is described by Figure 7.4, with the dashed rounded rectangle indicating Beauty's inability to distinguish the three tree positions.

A paradox arises from the two most popular given answers of 1/2 (corresponding to an expected value of $150) or 1/3 (corresponding to an expected value of $166\frac{2}{3}$). A brief argument for each position is given next.

As noted by Elga, the exact nature of the drug used to induce the sleep and amnesia is irrelevant. It only matters that Beauty *believes* she has no recall of the day of the week or a possible preceding interview. Of interest, though, is the fact that there are fast acting sedative medications, such as Versed® (midazolam), which are known to impair short-term memory. Some patients medicated with Versed have reported being "out" when post-surgically interviewed, despite the fact they were awake throughout the entire procedure in "conscious sedation."

The Halfers' Position

The probability of the coin having come up heads is obviously 1/2. It's a coin! Before Beauty is put to sleep she knows the probability of the coin coming up heads is 1/2. This should not change, unless she receives new and significant information. When awakened, she does not know if it is Monday or Tuesday and does not know if she has been previously interviewed; so, she has no new information and the probability should remain 1/2. If the experiment were to be performed many times, we should expect that roughly half the time the coin would come up heads. The day of the week and the number of interviews are irrelevant, and Beauty need not be concerned with her lack of recall. She is simply asked how likely it is that a single, fair coin has come up heads and the answer is clearly 1/2. As a result, her expectation at the conclusion of the experiment is $E = \frac{1}{2}(\$100) + \frac{1}{2}(\$200) = \$150$.

The Thirders' Position

During each interview, Beauty knows she is currently in one of three indistinguishable states:

1. The coin came up heads and it is Monday.
2. The coin came up tails and it is Monday.
3. The coin came up tails and it is Tuesday.

If the experiment were performed a large number of times, each state above would correspond to roughly one third of all interviews. Beauty knows she is being interviewed, but nothing more. She must assume the probability of state 1 above (heads) is 1/3 and expects to collect $E = \frac{1}{3}(\$100) + \frac{2}{3}(\$200) = \$166\frac{2}{3}$ when the experiment ends.

Halfers counter that "one out of three" does not correspond to a probability of 1/3. A halfer believes the three states are not equiprobable. Since the coin comes up heads with probability 1/2, state 1, the only state associated with heads, should also be assigned a probability of 1/2. States 2 and 3 should each be assigned a probability of 1/4. A thirder counters with the fact that one third of all interviews (indistinguishable to Beauty) correspond to state 1 and therefore the probability, which by definition is an event's theoretical relative frequency of occurrence, should be 1/3.

Halfers believe Beauty should estimate the theoretical relative frequency of all coin tosses that come up heads (1/2). Thirders believe Beauty should estimate the theoretical relative frequency of all interviews that correspond to heads (1/3). Both positions can't be correct; yet, both arguments seem airtight.

Are you a halfer or a thirder on this issue? If you were in Beauty's position and just awakened, would you accept $160 in exchange for what you would otherwise collect at the end of the experiment? If you are a halfer, you would take the offer because it exceeds your otherwise expected payoff of $150. If you are a thirder, you would decline the offer as you expect to collect 166\frac{2}{3}$ at the experiments conclusion.

The Ambiguity Position

A probability is a theoretical relative frequency of occurrence. But relative to what? Each experiment involves one toss of a fair coin. Before the experiment and during each waking, Beauty understands that each experiment is associated with one coin toss. So, relative to the total number of experiments, Beauty should expect heads to come up, on average, one out of two times and, based on this analysis, her credence in the coin having come up heads is 1/2. But, if relative frequency of occurrence refers to the total number of interviews, with all interviews given equal weight, then one out of three interviews corresponds to heads and Beauty will believe the probability is 1/3. Since *relative frequency of occurrence* is not clarified in the statement of the problem, the question is ambiguous.

This may be the most plausible explanation; yet, the question of Beauty selling her interest in the payoff remains. If during one of the indistinguishable wakings Beauty is offered $160 in exchange for her payoff at the end of the experiment, should she take it? If you were in Beauty's position, would you take it?

Applications

Applications are of two types, depending on whether we view imperfect recall as an impairment or an advantage. Impairment applications are not contrivances. We predictably forget things—names, phone numbers, passwords, etc. If you stop reading now and plan on resuming later, you will need a bookmark to find your place. Lack of recall is intensified by technological innovation that increases information access. We use "bookmarks" to recall websites and almost none of us can remember telephone numbers and addresses of acquaintances saved in handy electronic devices. Imperfect recall is the norm and is now a common component of memory models.

Mental Accounting

Itzhak Gilboa and Eva Gilboa-Schechtman draw parallels between the absentminded driver and consumers who do not keep perfect records of past spending [Gilboa and Gilboa-Schechtman 03]. A woman on a tight budget drives the freeway to work; but, she has the option of taking a toll road for a fee of $5. The toll road is faster, but the fee is high. She knows she will not be able to recall how many times she has taken the toll road since her last paycheck and is justifiably concerned that the fees could exceed her budget. Aware of her failing memory and knowing she can't afford to take the toll road every day, she adopts the pure strategy of never taking the toll road. This is analogous to the absentminded driver adopting the *ex ante* pure strategy of going straight at each possible exit, so as to avoid the hazardous exit . Clearly, she is over generalizing in assuming that if she were to take the toll road once, she would always take it. But imperfect recall forces the decision upon her. Though optimal at the planning stage, this strategy is suboptimal at the action stage. If she decides to skip the toll road all the time, then surely she could afford to take it just this one time. A way out of the dilemma would be some form of mental accounting that employs memory aids, but this violates the imperfect recall assumption.

Ameriks, Caplin, and Leahy consider absentmindedness and the relationship between spending and credit-card use—and linkages between financial planning and wealth accumulation and consumption at retirement [Ameriks, Caplin, and Leahy 04]. The model suggests a link between credit-card use and overspending. (Are we surprised?) Cash serves as a memory aid, in the sense that leftover cash in a consumer's wallet is an indicator of past spending. On the other hand, credit-card use promotes imperfect recall.

Texas Hold'em

Poker is a metaphor for life. You play the hand you're dealt and must know when to bluff, hold, or fold. A knowledge of the rules, patience, and recall of significant game events are as essential to poker as they are to life itself. Good luck helps as well. The game is relevant to the way we interact with others, conduct business, and plan for the future. A player's personality, be it defensive, aggressive, or unpredictable is reflected in the way he or she plays the game. We learn about ourselves when playing poker, and the game has been used as a teaching tool from grade school through college. As of this writing, Texas Hold'em is the world's most popular version of poker. Casinos offer live play, and online versions are available to anyone with an Internet connection. Hollywood movies and television coverage of tournaments have fueled the popularity.

It's no wonder that sociologists, psychologists, and computer scientists have developed models for discovering optimal play strategies. The Association for the Advancement of Artificial Intelligence (AAAI) is a nonprofit scientific society devoted to advancing intelligent behavior in machines and improving AI education. In 2006 they began holding the Annual Computer Poker Competition (ACPC), showcasing state-of-the-art programs for Texas Hold'em. Currently this annual event features programs submitted from over 30 countries. In many ways this parallels the development of chess-playing programs with some significant differences. In chess, the objective is to win. In poker the objective is to maximize winnings. Another key difference here is that chess is a game of perfect information whereas poker is a game of imperfect information (hidden cards). And, surprisingly, a higher degree of imperfect recall has proven to be advantageous in a recent ACPC tournament.

Before going into details, note that imperfect recall is not always an impairment. Indeed, selective memory (intentional amnesia) may be a good thing. As noted throughout this book, sometimes less is more (more or less). Under certain conditions, restricting information can be beneficial. Racehorses are equipped with blinders (blinkers, winkers) placed on the sides of their eyes to restrict their vision to the side and rear. Restricting visual information forces the horse to focus straight ahead on the task at hand and reduces the risk of distractions. Similarly, computer models of poker strategies may improve performance by selectively forgetting certain past actions of players and treating certain histories as being indistinguishable.

The extensive (decision tree) form of a perfect recall, heads-up (two-person) Texas Hold'em game, with limits on betting, is indeed *extensive*, having 10^{18} game states requiring petabytes of disk memory to store just one strategy. One petabyte is approximately 1,000 terabytes or 10^{15} bytes. To put this into perspective, this manuscript is being composed on a desktop computer equipped with a 160 gigabyte hard drive. One petabyte of hard drive storage is the equivalent of roughly 7,000 of these computers. An mp3 player with a petabyte of storage would hold approximately 300 million tunes or 2,000 years of music. In two-player, no-limit Texas Hold'em, the number of game states increases to 10^{71}, well beyond today's computers. Abstraction techniques that restrict the amount of information available to a player at a decision point reduce the size of the original game, producing an abstract game that hopefully models the original version.

Researchers from the University of Alberta and Yahoo! Research have used imperfect recall to produce the abstractions, and they claim their programs are stronger than the perfect-recall counterparts [Waugh et al. 09]. Using imperfect recall allows for more flexibility in designing the abstractions. Imperfect recall allows for more detailed information at early decisions, which may be discarded when not necessary at a later decision point. The abstractions are constructed in such a way that not all actions are fully observable to each player and certain past histories become indistinguishable. The imperfect-recall poker program submitted by Waugh et al. at the 2008 AAAI Computer Poker Competition won both the limit and no-limit events.

Modeling games and other decision processes must make extensive use of imperfect-recall abstractions simply because humans and computers are imperfect by nature. Technological advancement and the exponential growth of information require the sacrifice of irrelevant information to make good use of what really matters.

Non-zero-sum Games:
The Inadequacy of Individual Rationality

I call it the madman theory, Bob. I want the North Vietnamese to believe that I've reached the point that I might do anything to stop the war. We'll just slip the word to them that "for God's sake, you know Nixon is obsessed about communism. We can't restrain him when he's angry—and he has his hand on the nuclear button"—and Ho Chi Minh himself will be in Paris in two days begging for peace.
—Richard Nixon to Robert Haldeman

A s discussed in Chapter 3, the fundamental theorem of game theory (or minimax theorem) guarantees the solvability of two-person zero-sum matrix games. That is, for every such game, there exists optimal strategies for each player and an associated expected payoff or value of the game. Rational players should employ the optimal strategies, and the long-term expected payoff is predetermined. Solutions to such games are universally accepted.

Zero-sum refers to a player's win being accounted for by the other player's loss. When considered collectively, there is zero collective gain for both players after each play. There is no sense of cooperation and there would be no point in preplay communication where the players discuss potential strategies. It's simply a matter of kill or be killed. There is no win-win.

Non-zero-sum games need not require a zero-sum gain for both players. In these games, both players can win, both can lose, and a player's gain need not equal the other player's loss. These games are not strictly competitive; cooperation may be desirable. The air traffic controller and the pilot play (perhaps a poor choice of words) so as to safely land the airplane. Both are satisfied when all goes well. If played poorly by both, each experiences a horrific loss. Neither thinks in terms of "beating the other." So, it's like a dance where the players do their best, yet cooperate as necessary for an optimal outcome for each.

Two children tossing a Frisbee back and forth in the park are playing a non-zero-sum game. Would it make sense to ask who is winning? Of course not! Other examples include social interaction, labor management negotiations, and military conflict.

Non-zero-sum games differ from zero-sum games in more ways than definition alone. There is no general solvability of non-zero-sum games as is the case with zero-sum games. Preplay communication and negotiation often leads to payoffs for both players better than those if no communication existed. Again, such a concept is nonexistent for zero-sum games. Most striking is that individual rationality, desirable for successful play of zero-sum games, often proves inadequate for optimal play of non-zero-sum games. In fact, a relatively rational player may be at a disadvantage when paired with an irrational player!

This chapter includes several classic examples, most all of which have been played by the reader in one form or another.

Pizza or Pâté

To introduce the concept, terminology, and some basic notation, consider friends Ron and Carla deciding where to go for dinner one evening. Being friends and enjoying each other's company, they prefer to dine together rather than at separate restaurants. On this occasion Ron prefers a casual pizza-and-beer establishment but Carla would much rather sample the pâté at the elegant Chez François. Both would rather dine together at either restaurant than dine alone.

As given in Matrix 8.1, payoffs are of the form (r,c) where r denotes Ron's payoff and c denotes Carla's payoff. The payoffs must appear as ordered pairs because Ron's win need not equal Carla's loss and both quantities must be specified. For each player a positive number denotes a win and a negative number represents a loss. It's a non-zero-sum game and, in general, $r + c \neq 0$.

We should interpret the payoffs as follows: Both Ron and Carla are relatively content to dine together, although Ron would find it more enjoyable going out for pizza and Carla would prefer if they dine at Chez François. Both are equally un-

	Carla	
	Pizza	**Pâté**
Ron — **Pizza**	(2,1)	(−1,−1)
Ron — **Pâté**	(−1,−1)	(1,2)

Matrix 8.1. Pizza or pâté payoff matrix.

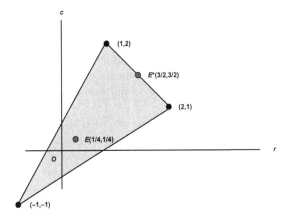

Figure 8.1. Pizza or pâté payoff polygon.

satisfied to dine at separate restaurants. Envy is not a consideration. Ron prefers dining at the French restaurant with Carla to dining alone, despite the fact that Carla will enjoy the experience more than he.

The payoffs (ordered pairs) in Matrix 8.1 can be displayed graphically to form a *payoff polygon* or *convex hull*, the set of points that is the smallest convex polygon containing all payoff points. Picture the payoff polygon as being formed by stretching an elastic band around all payoff points. It will shrink to form the payoff polygon.

The shaded interior plus the boundary represent all possible expected payoffs associated with pure, mixed, and coordinated strategies. For example, if Ron and Carla each toss a coin to determine at which restaurant they will respectively dine, then the expected payoff is

$$E = \frac{1}{2}(-1,-1) + \frac{1}{4}(1,2) + \frac{1}{4}(2,1) = \left(\frac{1}{4}, \frac{1}{4}\right)$$

and is denoted by point E in the polygon's interior. Clearly, they both can do better as other strategies yield points up and to the right of E.

What strategies should Ron and Carla adopt so as to maximize their payoffs? Is this game solvable? If so, in what sense? The fundamental theorem of game theory does not apply to non-zero-sum games, and optimal strategies are not clear, if indeed they do exist.

A casual glance at Figure 8.1 suggests value in preplay negotiations. The game is symmetric, so an equitable solution would correspond to $(3/2,3/2)$, denoted as E^*, on the upper-right edge. This is optimal in the sense that both players can't get equal payoffs of greater value, no matter how they play. This expected payoff is intuitively achieved by coordinated compromise. Ron and Carla might agree to alternate their dinner dates by going together for pizza one evening, then together for pâté the next evening, and so on. Or, a single coin could be tossed to determine at which restaurant they mutually dine. Such preplay negotiations are meaningless when it comes to zero-sum games. Can we declare the game to be *solved* under the assumption that preplay communication is allowed? The unexpected answer is no.

What if Ron is irrationally stubborn, so much so that he moves first by making a reservation and nonrefundable payment for the evening's dinner at the pizza parlor? Ron tells Carla, "We have no choice. I've bought the dinners and I'm going, with or without you." Though irrational, this stubbornness pays off for Ron. Ron's stubbornness wins out over Carla's rational willingness to compromise. Carla has no other option than to "cave in" and join Ron at the pizza parlor. From Ron's point of view, this is better than the rational compromise because his payoff is now 2 as opposed to 3/2. And Carla's payoff is worse than that of a compromise, yielding her 1 as opposed to 3/2. The pizza-pâté game is a battle of will, not wit, where irrational stubbornness prevails over rationality. Ultimately, there is no general solution to this game as there is with zero-sum games.

This game is well known in game theoretic literature. Luce and Raiffa call it "Battle of the Sexes" [Luce and Raiffa 85, pp. 90–94]. Consider the contrast between zero-sum games and non-zero-sum games:

1. All two-person zero-sum games as given in Chapter 3 are solvable, having universally accepted solutions. This is not so with non-zero-sum games, which are more applicable to day-to-day life than zero-sum games.
2. It is desirable to play rationally when playing zero-sum games. This is not always the case with non-zero-sum games, as the previous example illustrates.
3. It is never advantageous to move first in a zero-sum game. Doing so only gives the opponent additional information to be exploited. It may be advantageous to move first in certain non-zero-sum games, as the above example illustrates.
4. For zero-sum games, the concept of preplay communication (or negotiations) is meaningless. The above example suggests such communication is of potential value with respect to certain non-zero-sum games.

The pizza-pâté game, though contrived, models social interactions we've all experienced. Have you ever found yourself dealing with someone so stubborn that you found it pointless to try and compromise? Perhaps under those circumstances you decided to yield, giving the other person greater satisfaction (a larger payoff) than yours.

The Threat

The previous example shows there may be a psychological component when it comes to optimal strategies for non-zero-sum games. Preplay communication, negotiations, and even irrational play may be used by one or both players to increase their respective expected payoffs. As a result, non-zero-sum games are used to model a variety of social interactions, and the literature is rich with examples, definitions of solvability, and optimal strategies and applications. Labor-management bargaining, hostage-release negotiations, and estate settlement are typical real-world scenarios to which this theory may be applied.

Non-zero-sum games are not generally solvable as are zero-sum games. But, one would suspect that preplay communication may be of value. In the previous example, Ron and Carla may agree to cooperate giving an expected payoff of (3/2,3/2). This payoff could never be achieved without preplay communication as random play without coordination yields the payoff (1/4,1/4), an inferior result for both players. Even if Ron stubbornly insists on pizza and "moves first," the payoff (2,1) remains superior to (1/4,1/4) for both players. One is tempted to draw the conclusion that no harm can come from preplay communication and there is potential benefit for both players. After all, what harm can come from negotiating? In fact, there can be harm and, under certain circumstances, it would behoove a player not to negotiate and remain *incommunicado* prior to play. Luce and Raiffa offer the following example [Luce and Raiffa 89, p.111].

Without preplay communication and assuming the game is to be played only once, optimal strategies for the game shown in Matrix 8.2 are clear. For Ron, row 1 dominates row 2 in the sense that no matter which column Carla chooses, Ron is better off choosing row 1 rather than row 2. Similarly, Carla observes the domination of column 1 over column 2 because she is better off choosing column 1 no matter what Ron decides to do. This is essentially the same as the row and column domination discussed in Chapter 3. Assuming both Ron and Carla play rationally, they will choose row 1 and column 1 respectively, yielding a payoff of (1,2). The game appears strictly determined with pure optimal strategies and hence solved.

Carla

	Column 1	Column 2
Row 1	(1,2)	(3,1)
Row 2	(0,−200)	(2,−300)

(Ron labels the rows)

Matrix 8.2. The threat.

Things get interesting if we allow the players to communicate prior to play. Ron, with a previously established reputation of being stubborn, irrationally insists that Carla play column 2 instead of the dominating column 1. Carla questions Ron, asking him why he demands this. His response, "Because I can! If you don't do as I ask, I will play row 2 instead of row 1. Yes, my payoff will drop slightly from 1 to 0. But it will hurt you much more than it will hurt me. Your payoff will drop from a win of 2 to a loss of 200. You don't want this to happen, do you? Then do as I say."

Carla should find the threat credible as Ron has already established himself as being a bit irrational in previous game play. Having reason to believe he will carry out his threat, she agrees to play column 2. Ron sticks with row 1 and the payoff is now (3,1), giving Ron his best possible payoff and Carla her second best. She has reason to regret the preplay communication and by now probably regrets ever meeting Ron in the first place. Despite her disgust, Carla plays the game and wins 1 unit, a slightly better result than walking away without playing. Ron smirks and collects his winnings of 3. Carla says to Ron, "You're a pig, a selfish pig! You have no sense of fairness". Ron replies, "I'm just trying to do the right thing—for me." Carla breaks off the relationship.

The point is that preplay communication is of no benefit to Carla; indeed, it may do more harm than good. Without communication there could be no treat. Carla should have played the game without preplay discussions, letting the doorbell ring and not returning Ron's calls.

Even without direct preplay communication, Ron can coerce Carla into playing column 2 with repeated plays of the game. Ron's threat could be signaled by his playing row 2 until Carla submits. Ron's communication through iteration is as effective as direct communication though may take longer to achieve the desired effect.

Chicken: The *Mamihlapinatapei* Experience

The non-zero-sum game *chicken* is surely one of game theory's most analyzed games. The game's two players interact so as to be on a collision course with

	Carla	
	Compromise	No compromise
Ron Compromise	(0,0)	(–1,1)
No compromise	(1,–1)	(–10,–10)

Matrix 8.3. Chicken.

catastrophic consequences to both if neither player compromises. The preferred outcome for each is to hold the course with the other player compromising. A less desirable outcome, though far from catastrophic, would be for both players to compromise. The payoff matrix is given in Matrix 8.3.

Chicken is a game of brinksmanship, where players hold their course to the brink of disaster in the hope of achieving some goal. In the 1955 James Dean movie *Rebel Without a Cause*, Dean's character Jim is challenged by Buzz to a "chickie run." Jim and Buzz each race stolen cars to an abyss, and the first to jump out before the car goes over the cliff is branded a chicken and loses status. Buzz gets a strap on his leather jacket caught on a door handle and becomes stuck in the car as it goes over the cliff. Jim survives, by jumping out before his car goes over. Fortunately, the game never became popular beyond Hollywood films about juvenile delinquents.

The name *chicken* is now applied to a broad category of such games, possibly as a result of Bertrand Russell's comments comparing nuclear brinkmanship to a form of highway chicken [Russell 59, p. 30]:

> Since the nuclear stalemate became apparent, the Governments of East and West have adopted the policy which Mr. Dulles calls "'brinkmanship'". This is a policy adapted from a sport which, I am told, is practiced by some youthful degenerates. This sport is called "Chicken!". It is played by choosing a long straight road with a white line down the middle and starting two very fast cars towards each other from opposite ends. Each car is expected to keep the wheels of one side on the white line. As they approach each other, mutual destruction becomes more and more imminent. If one of them swerves from the white line before the other, the other, as he passes, shouts "Chicken!", and the one who has swerved becomes an object of contempt. As played by irresponsible boys, this game is considered decadent and immoral, though only the lives of the players are risked. But when the game is played by eminent statesmen, who risk not only their own lives but those of many hundreds of millions of human beings, it is thought on both sides that the statesmen on one side are displaying a high degree of wisdom and courage, and only the statesmen on the other side are reprehensible. This, of course, is absurd. Both are to blame for playing such an incredibly dangerous game. The game may be played without misfortune a few times, but sooner or later it will come to be felt that loss of face is more dreadful than nuclear annihilation. The moment will come when neither side can face the derisive cry of "Chicken!" from the other side. When that moment is come, the statesmen of both sides will plunge the world into destruction.

A quick check of Matrix 8.3 shows no row or column domination; so, there are no obvious pure strategies for either player. What is clear is that each player prefers to do the opposite of the other player. The problem, of course, is that neither player knows with certainty what the other player will do.

Rather than use the highway version of chicken, let's consider a more contrived but illuminating example. Ron and Carla agree to be placed inside the "chicken chamber," an airtight gas-chamber enclosure with two seats. The door will be closed and the chamber sealed. Both understand that at some time within the next hour a cyanide capsule will be released into the chamber, instantly killing them both, unless prior to the capsule's release one of them presses a panic button. If either presses this button prior to the release of the capsule, the door to the chamber will immediately open and the capsule will not be released, at which point in time the two will be free to exit. Pressing the button is to be thought of as compromising, analogous to swerving in highway chicken. Not pressing the button is holding the course. What should they do? A better question is to ask, why they have agreed to play this game in the first place? And what's with Carla? Why is she choosing to be in the same room with Ron, let alone a potential death chamber? For our purposes consider the payoffs in Matrix 8.3 to be millions of dollars. So, if Ron holds the course and Carla hits the panic button, Ron wins $1 million and Carla loses the same. In the event neither compromises, they both lose $10 million, which is being equated to instant death. (This may not be an accurate comparison, although some individuals have committed suicide after losing lesser amounts.)

If the players are not allowed to communicate and both choose their options randomly (compromising with probability one-half), the expected payoff to each is given by

$$E = \frac{1}{4}(0,0) + \frac{1}{4}(-1,1) + \frac{1}{4}(1,-1) + \frac{1}{4}(-10,-10) = \left(-\frac{5}{2}, -\frac{5}{2}\right),$$

a dismal outcome for both. Each might reason, "I don't want to die. I must compromise immediately so as to avoid certain death." Assuming both reason cautiously, the payoff to each is zero. If we assume Ron and Carla are rational, at least to the extent they wish to avoid death, then it's a safe bet they will both compromise. But, each player will assume the other will almost certainly compromise. If the other player is certain to compromise, then why not hold the course and win the million? No solution!

The symmetry of the matrix suggests an equitable agreement is achievable by preplay negotiations. The preferable solution would be a binding agreement where they both promise to compromise. Play would begin, they would both hit the panic button, and the game would end with a zero payoff to each player.

But what if preplay communication is not allowed? Perhaps Carla refuses to negotiate with Ron, considering what has transpired when playing the previous two games. It's game time and both enter the chamber, neither speaking to the other. After being seated and the door closes, Carla smells alcohol in the chamber. She glances over at Ron and, to her horror, notes his bloodshot eyes. He's drunk! He loses his balance and collapses to the floor. The chamber is sealed. Carla now realizes Ron is incapable of hitting the panic button and that she must do it or else die. She does and the game ends with Ron winning $1 million and Carla phoning her lawyer and financial advisor.

As is the case with the pizza-pâté game, it is advantageous for Ron to move first and let Carla deal with it. This is equivalent to Carla being able to read Ron's mind and knowing his move before he actually makes it. (Carla may know Ron to be an alcoholic.) Mind reading is advantageous to players of zero-sum games. For the pizza-pâté game and chicken, it works against the omniscient player, creating a *paradox of omniscience.*

Chicken, perhaps more than any other non-zero-sum games, clearly illustrates the inadequacy of individual rationality. More than being inadequate, Ron's lack of awareness virtually guaranteed his win and Carla's loss. Rationality, a highly desirable quality for players of zero-sum games, may be undesirable for certain non-zero-sum games. And, the more irrational and less physically capable, the better. Ron might as well have shown up unconscious with two broken arms. His inability to play, mentally or physically, guarantees him the best possible payoff with Carla surviving but suffering a significant financial loss. This is reminiscent of the Chapter 6 truelers. Survival of the weakest? Well, it's more like survival of the totally incapacitated.

In this regard there is something both paradoxical and troublesome about all sorts of social, political, and military interactions. Intelligence (rationality) may not be as much of an asset as a liability. To be more precise, it is the perception of irrationality, not irrationality itself, which may favor the irrationally appearing player. Once Carla perceived Ron as being mentally and physically incapable of compromising, she had no choice but to do so herself. From this chapter's epigraph it's clear that Richard Nixon attempted to apply this concept when dealing with Ho Chi Minh and the North Vietnamese. In this case the strategy failed. It's frightening to think that there may be times in our future where world leaders will behave irrationally (in actuality or perceived) and consequently may get their way precisely because of the irrational behavior. Even more troublesome is the fact that both sides can employ the same strategy.

Recent history is rich with examples of military, foreign relations, and political brinkmanship, most notable of which is the Cuban missile crisis of 1962, the most serious confrontation of the Cold War. The brief, sixteen-day confrontation brought the United States and the Soviet Union perilously close to war and may have ended in nuclear war had no compromise been reached.

Prior to the crisis, the Soviets had begun placing intermediate-range nuclear missiles in Cuba to counter US strategic capabilities and protect Cuba against an invasion similar to that of the failed Bay of Pigs fiasco of 1961. Evidence of the nuclear missile installations emerged in the summer of 1962, when high altitude pictures taken by US spy planes confirmed their existence. President Kennedy wanted the missiles removed and the Soviets, under Premier Nikita Khrushchev, refused. A game of chicken had now begun.

On October 22, President Kennedy addressed the nation and the world, announcing the discovery of the installations and warning that a missile attack on the US from Cuba would be viewed as an attack by the Soviet Union and would be followed by an immediate and appropriate US response. He announced a naval blockade of Cuba to prevent Soviet ships bringing additional missiles and associated construction equipment to the island. Khrushchev's reply was a stern warning that should the Americans interfere with any Soviet ship, the USSR would be forced to take all necessary action.

Tensions rose as the game played out with the United States, the Soviet Union, and Cuba preparing for the possibility of nuclear war. Behind the scenes, negotiations were taking place in the hope of reaching some sort of compromise and averting disaster. Letters were exchanged between Kennedy and Khrushchev. Even Bertrand Russell entered the fray, sending telegrams to world leaders including Secretary-General of the United Nations U Thant.

On October 26, a compromise was reached and the crises effectively came to an end. The Soviets agreed to dismantle the installations in Cuba and return the weaponry to the USSR. The United States agreed to end the quarantine with additional assurances to not invade Cuba.

Budget negotiation deadlocks at the federal and state levels are modeled similarly. For financial and political reasons, neither side may be willing to compromise with the potential catastrophic result of no budget and the government shutting down. A classic example of economic/political chicken is that of the 2011 US debt ceiling crisis. On August 2, 2011, after Congress approved a plan to raise the debt ceiling, President Obama signed into law the Budget Control Act, bringing to an end an agonizingly long and painful game of political and economic chicken played by US lawmakers. The agreement delayed, at least temporarily, the potential economic disaster of the US defaulting on its legal obligations to pay its debt.

It's a matter of opinion when the crisis began. Since 1917 the US Congress routinely raised the debt ceiling limit to accommodate the government's need to borrow. Until 2011 the act of raising the limit was a matter of formality, with little opposition, carried out whenever there was a need to borrow in excess of the current limit. In 2010 the debt ceiling limit was raised to $14.294 trillion. On May 16, 2011, this limit was reached but Congress refused to raise the limit and the US Treasury stopped borrowing money. In 2011 raising the limit was far from a formality: US opinion and the political makeup of Congress was highly polarized. Republicans opposed increasing taxes to reduce the debt, favored broad cuts in federal spending, and were opposed to raising the debt limit unless accompanied by severe cuts in spending with no increase in taxes. Democrats favored increasing taxes, while opposing cuts to Medicare and other social welfare programs.

Moving first behavior, shown by both political parties, took the form of ideological pledges, or vows, to hold fast to a stated position and fail to yield, no matter what. This significantly delayed compromise, pushing the crisis to the brink of economic disaster. In many cases pledges were made to secure funding from special interest groups. As strategic moves in chicken, the pledges were preemptive, signaling to the opposition an unwillingness to compromise.

The Taxpayer Protection Pledge

The Americans for Tax Reform collected signatures for a pledge opposing all tax increases and the elimination of tax breaks. As of 2011, over 270 members of Congress had signed the pledge.

The Cut, Cap, and Balance Pledge

House Republicans describe Cut, Cap, and Balance as a "common-sense plan to rein in the debt":

Cut. Substantial cuts in spending that will reduce the deficit next year and thereafter.

Cap. Enforceable spending caps that will put federal spending on a path to a balanced budget.

Balance. Congressional passage of a Balanced Budget Amendment to the US Constitution—meaning an end to borrowing—but only if it includes both a spending limitation and a super-majority for raising taxes, in addition to balancing revenues and expenses.

The Social Security Protection Pledge

One hundred twenty-three Democratic members of Congress pledged to defend Social Security from privatization, block any increase in the retirement age, and oppose cuts in benefits.

On August 2, 2011, a compromise was reached, raising the ceiling by $2.4 trillion and providing just under $1 trillion in spending cuts over 10 years. Despite the eleventh-hour Budget Control Act compromise, Democrats, Republicans, the President, and the US economy all took hits. Congress' polarization and the resulting gridlock was followed by historically low public approval, with only 6 percent of Americans believing Congress was doing good or excellent work; some called it the "worst Congress ever." Most believed lawmakers were more interested in promoting their own careers than helping the country. President Obama's approval ratings dropped, with members of both parties believing he showed a lack of leadership on the crisis. On August 5, 2011, only three days after the debt ceiling bill was signed into law, the rating agency Standard & Poor's Ratings Services pessimistically downgraded the US credit rating from AAA to AA+. This triggered a significant drop in major stock market indices around the world.

Labor-management negotiations are similar, with no compromise leading to a strike resulting in lost wages and loss of profit. Consumers will lose the goods and services. In the event emergency services are involved, the loss would not be well tolerated. Police negotiations with a hostage taker or terrorist exhibit the characteristics of chicken. If neither side is willing to compromise, there may be loss of life as might occur in a botched rescue attempt.

If there are three or more players, the game is a *volunteer's dilemma*, where any one player may take a compromising action to benefit all. Picture the unpleasant scenario of being one of twenty passengers stuck on a lifeboat in freezing ocean waters. Bad weather suggests no rescue ship or aircraft will reach you for at least one day. Survival for the next 24 hours would be no problem, except for the fact your lifeboat is overloaded by 100 pounds and is taking on water. It will surely sink and all will die unless one of the 20 passengers volunteers to leave the lifeboat and go into the frigid waters. The volunteer will die, but in doing so he or she will save the lives of the other 19 passengers. If nobody volunteers then all will die. What should be done? Mathematics is at a total loss when it comes to such matters.

A US infantry manual published during World War II addressed the problem of a live grenade falling into a trench in which soldiers were sitting. A soldier was advised to place himself over the grenade, sacrificing himself to save others. But which soldier should volunteer? And, the decision must be made within seconds.

A sad and well-known example of the volunteer's dilemma is that of Catherine "Kitty" Genovese of Queens, New York, murdered on March 13, 1964, while walking home to her courtyard apartment. At approximately 3:20 am she was stabbed multiple times by an assailant. The attack, lasting over 30 minutes, was witnessed by 38 neighbors from their courtyard balconies as she cried out for help. None of the neighbors came to her immediate aid or phoned the police during the assault. The police received their first call at 3:50 am and arrived within a few minutes. At 4:25 am an ambulance arrived to take away the deceased victim.

Neighbors were questioned as to why they did nothing during the attack. Some admitted to being afraid, some simply didn't want to get involved. A volunteer's dilemma attitude may have been a contributing factor. Peering down from their balconies witnesses may have thought, "Someone will definitely call the police. There is no need for me to phone as the call has probably been made. There is no need for me to get involved and I'll just stay out of it." All knew that someone had to do something; all assumed someone else would do it.

Six days later, Winston Mosely was arrested and charged with Genovese's murder.

Social psychologists now describe such reluctance to get involved as the *Genovese* or *bystander effect*. Soon after this incident the 911 emergency telephone system was introduced in the US, allowing for the anonymous reporting of a crime.

James Patrick Kinney's beautiful poem, *The Cold Within*, is about lack of tolerance and discrimination. A volunteer's dilemma, mixed with racial, religious, and ideological hatred, can only end sadly.

> Six humans trapped by happenstance,
> In black and bitter cold.
> Each one possessed a stick of wood,
> Or so the story's told.
>
> Their dying fire in need of logs,
> The first woman held hers back,
> For on the faces around the fire,
> She noticed one was black.
>
> The next man looking 'cross the way
> Saw one not of his church,
> And couldn't bring himself to give
> The fire his stick of birch.
>
> The third one sat in tattered clothes;
> He gave his coat a hitch.
> Why should his log be put to use
> To warm the idle rich?
>
> The rich man just sat back and thought
> Of the wealth he had in store,
> And how to keep what he had earned
> From the lazy, shiftless poor.
>
> The black man's face bespoke revenge
> As the fire passed from his sight,
> For all he saw in his stick of wood
> Was a chance to spite the white.
>
> And the last man of this forlorn group
> Did naught, except for gain.
> Giving only to those who gave,
> Was how he played the game.
>
> The logs held tight in death's still hands
> Was proof of human sin.
> They didn't die from the cold without.
> They died from the cold within.

The French have a word for almost everything; but here the credit goes to the Yaghan of Tierra del Fuego. The word *mamihlapinatapei* loosely translates as "looking at each other hoping that either will offer to do something that both parties desire but are unwilling to do." The same word applies to awkward social situations where one party wishes the other would initiate a conversation or engagement but is unwilling to do so themselves. *The Guinness Book of World Records* describes it as the "most succinct word," and it's considered one of the hardest words to translate. Think of it as "eye contact suggesting 'you go first.'"

Ronny Cox, the actor who plays Drew in the 1972 movie *Deliverance*, is now a singer/songwriter. His recording *Mamihlapinatapei* appears on the CD *Ronny Cox: Live*.

They both swear they met by chance
He liked the band, she liked to dance
They were in love at their first glance
Mamihlapinatapei

All that summer, no matter what they did
You could always see them with her hand in his
They almost said it once, but God forbid
Mamihlapinatapei

Chorus:
Funny how the patterns of our well-laid plans
Tangle our way
Scared that we'd unravel if we said the words
We never say Mamihlapinatapei

We can't say what's in our hearts
Our minds keep saying "Is this smart?"
But our eyes ask "Can't we start?"
Mamihlapinatapei

Chorus

They were married yesterday
Two lovers danced the night away
Words could never touch what their eyes convey
No, words can't touch what their eyes say
Mamihlapinatapei....Mamihlapinatapei.....Mamihlapinatapei

The Prisoner's Dilemma

The 1998 film *Return to Paradise* is the story of three Americans, Lewis, Sheriff, and Tony, traveling together around Malaysia, enjoying the beach, girls, and hashish. Sheriff and Tony return to the states while Lewis stays on in Malaysia with the leftover hashish. Soon after, Malaysian authorities discover a significant amount of the hashish in Lewis' possession and he is arrested. Being over the legal limit for possession by one person, trafficking is presumed. Lewis is convicted and sentenced to death.

Two years later, Sheriff and Tony are contacted by Lewis' lawyer, Beth. She gives them the bad news about Lewis' conviction and that the execution is scheduled in eight days. She tells them an agreement was made with the Malaysian authorities. If Sheriff and Tony return to Malaysia and confess to their ownership of the hashish, Lewis will be allowed to live. Sheriff and Tony will each spend three years in prison. If only one of them returns, he must serve six years.

Of course, neither Sheriff nor Tony wants Lewis to die. But, the thought of serving three years in a Malaysian prison is not high on their "to do" list. Worse yet, there's the possibility of serving six years. It's a moral dilemma about legal ethics and personal responsibility.

This is Hollywood's version of the *prisoner's dilemma*, the most written-about non-zero-sum game in literature. Merrill Flood and Melvin Dresher, working for the RAND Corporation, devised the game in 1950. Innumerable articles appear in books, newspapers, academic journals, and popular science publications. It's even part of some British and US television game shows. A variety of real-world dilemma's can be modeled by the game.

Canadian-born American mathematician Albert Tucker, while working as a RAND consultant, gave the game its name, "prisoner's dilemma," based on his anecdote to illustrate it. In Tucker's formulation, imagine that you and an accomplice are arrested and charged for possession of a concealed weapon and armed robbery. There is enough evidence to convict you both on the possession charge, but the police are hoping to get a conviction on the more serious charge of armed robbery. You and your accomplice are held in separate cells and interrogated. You each must independently choose whether or not to confess to the armed robbery charge—and thus implicate the other—or stay silent. If both of you confess, each will be given a ten-year prison sentence. If neither of you confesses, both will be convicted of the lesser charge of possession and sentenced to one year each. The police offer both of you the deal that if one of you confesses and the other doesn't, then the confessor will be set free and the other will be convicted and sentenced to 20 years on the charge of armed robbery.

The payoff matrix is given in Matrix 8.4 where terms of a sentence are represented by negative numbers.

What is your choice if your only goal is to spend the least amount of time behind bars? We assume you have no relationship with your accomplice other than being partners in crime. Neither of you would lose a night's sleep if you never saw the other again.

At first glance, the rational strategy seems obvious. If your accomplice doesn't confess, you should, in which case you go free. Otherwise, you will spend a year in prison. If your accomplice confesses, then you should do so as well, resulting in ten years behind bars. Otherwise, you will spend 20 years behind bars. So either way, you are better off confessing. In game-theoretic terminology, row 2

Accomplice

	Don't confess	Confess
Don't confess	(–1,–1)	(–20,0)
Confess	(0,–20)	(–10,–10)

You

Matrix 8.4. Tucker's formulation of the prisoner's dilemma game.

dominates row 1 and you should certainly confess. Symmetry suggests your accomplice should do the same. No matter what you choose, your accomplice is better off confessing. Column 2 dominates column 1 in the same way as row 2 dominates row 1. By this line of reasoning and assuming both of you are acting rationally, you will both confess to the higher charge and each will be sentenced to ten years in prison. Optimal strategies are pure, and the game is strictly determined and hence solved. Right?

It's not so simple. Assuming you both do what you are rationally compelled to do, you will each spend ten years in prison. You both agree that a mutually superior outcome is for each of you to stay mum, resulting in a sentence of only one year each. You both know this and understand this is easily achieved if neither one of you confesses. Yet, individual rationality seems to prevent this possible payoff.

To try and get a handle on things, we reduce the game to simplest terms by removing the prisoner analogy and giving the payoffs as nonnegative integers. Ron and Carla play the game where the payoffs are given in dollars. If they both cooperate (analogous to each cooperating with their partner in crime and not confessing), each will win $2. If they both defect (analogous to each ratting on the other), each wins only $1. If one defects and the other cooperates, the defector gets the maximum possible payoff of $3 and the sucker gets nothing. (See Matrix 8.5.)

Carla

	Cooperate	Defect
Cooperate	($2,$2)	($0,$3)
Defect	($3,$0)	($1,$1)

Ron

Matrix 8.5. A simplified prisoner's dilemma game.

The game is symmetric and, for each player, defection dominates coopera-tion. Mutual defection generates a $1 payoff for each. Yet, they both could double their winnings by being nice and mutually cooperating. Why won't each decide to do this? Because Ron knows that no matter what Carla does, he is better off defecting. The same holds for Carla. Both understand that cheating (defecting) undermines the common good, yet they are unable to rationally avoid it. Imagine yourself as one of the players. What would you do?

As human beings we think of ourselves as rational, acting in our best inter-est whenever possible. But Ron and Carla, playing rationally as described, are not acting in their best interest. Hamsters could do better. It's true! (Let's assume hamsters would "choose" randomly and irrationally, having no way of compre-hending the nature of the game.) If our two hamsters are faced with the games two choices (meaningless to a hamster) and choose randomly, the expected pay-off is

$$E = \frac{1}{4}(\$2, \$2) + \frac{1}{4}(\$0, \$3) + \frac{1}{4}(\$3, \$0) + \frac{1}{4}(\$1, \$1) = (\$1.50, \$1.50),$$

a superior result for each than Ron and Carla's paltry ($1,$1). What's going on here?

Rational, mathematical logic is at a total loss when it comes to such games, and there is no generally accepted solution. One potential way out of the dilemma is for the players, in this case Ron and Carla, to meet before the game and ne-gotiate a binding agreement guaranteeing each will cooperate. Since there will be an incentive for each to break the agreement, it must be enforceable, with teeth, guaranteeing mutual cooperation and the ($2,$2) payoff. This may be the only way this payoff can be achieved by rational players on a single play of the game. Do rational beings require an overseeing authority to force them to act in their own best interest? The prisoner's dilemma game suggests this may be so. Of course, this overseeing authority may take the form of a personal code of ethics or belief in a god. (God will punish us if we don't play nice.) This inadequacy of in-dividual rationality goes against the grain of Adam Smith's invisible hand, which suggests that individuals acting on their own behalf are guided in such a way as to promote the public interest. Considerations given to the prisoner's dilemma force us to question whether self-serving decisions serve society as a whole. Here is a personal anecdote:

I regularly teach a course in finite mathematics at El Camino College and spend a day or two covering the prisoner's dilemma and other non-zero-sum games. I devised an experiment to see how my students would play the game. It took the form of a short, five-minute quiz, where my students were given the opportunity to earn extra-credit points proportional to the values given in Matrix 8.5. The points would then be added to their final exam score. Each student was given a blank sheet of paper, on which they were to write "cooperate" or "defect" to indicate their move in the game. The students were told that I would pair each of their papers with another randomly chosen student and that the extra credit points would be determined by how each student and the anonymous partner play. Students were reminded that they were not trying to *beat* the other player. They should simply try to secure as many extra credit points for themselves as possible. Preplay negotiations were impossible since no student knew the identity of their playing partner. I also required my students to give a brief justification of their choice. Since we discussed the game extensively before actually playing it, the class understood that it was in their best interest for all to cooperate.

Did it happen? Far from it! The tendency to defect was overwhelming as it was easy to do so anonymously. Of the 30 students participating, 28 defected and two cooperated. As it turned out, the two "suckers" were not paired with each other and therefore earned no extra-credit points. Everyone else in the class received minimal extra credit. They could have done better.

And why did the suckers cooperate? One admitted to not understanding the game. The other said she was afraid others might find out she defected and be upset with her. As it turned out, over 90 percent of the class defected. Once again, hamsters would have done better! The class might have preferred a proctor with the authority to walk around the room during the quiz, checking to make sure all cooperated.

The class wanted a second chance, an extra-credit re-test. Their request was denied; however, their request raises an interesting question. If a significantly large number of iterations are played, will cooperation evolve? Will there be communication and hence cooperation through iteration similar to what happens with The Threat? Robert Axelrod, a leading expert on game theory and Professor of Political Science and Public Policy at the University of Michigan, says "yes" [Axelrod 84, p. 173]. In 1979 he conducted a prisoner's dilemma computer tournament and discovered that cooperation does indeed evolve, through reciprocity. The most successful strategy, named Tit for Tat, was to cooperate on the first move and then do whatever the other player did on the previous move. Once cooperation got started, it was sustained by the players themselves without the need for direct communication between players or a governing authority.

For cooperation to evolve, it is required that the players meet a large and indefinite number of times. If two players understand they are to encounter each other a finite number of times, another paradox develops. Let's say Ron and Carla understand that they will meet and play the game ten times without direct communication. On the tenth and last play of the game, they both know there is no future, no reciprocity, and no reward for being nice. It's effectively a single play, and with no incentive to cooperate, we expect mutual defection. But, the same may very well occur on the ninth play of the game as they both know the predetermined outcome of the tenth. The argument extends all the way back to play one, suggesting a good possibility of mutual defection on all ten games. But, if the players do not know when the last game will take place, and if the number of games is sufficiently large, cooperation can, and probably will, evolve.

The entertainment value of the prisoner's dilemma game has not been lost on television game show developers. *Golden Balls*, a British daytime game show, includes a segment called "Split or Steal," where two contestants simultaneously decide whether to split or steal a jackpot worth as much as £100,000. They are allowed to discuss the dilemma briefly before making their moves simultaneously. If they both choose to split (cooperate), they each get 50 percent of the jackpot. If one chooses to split and the other chooses to steal (defect), the stealer gets 100 percent and the splitter gets nothing. If both choose to steal then neither player wins anything. During the brief preplay discussion, the players most always agree that splitting is best. But, temptation often causes one or both contestants to steal. Since any given pair of contestants play only once, there is no chance for cooperation through iteration.

The prisoner's dilemma has been a component of the television game shows *Shafted* (British) and *Friend or Foe* (US). The Australian morning radio show *Kyle and Jackie O* features a game where two callers are randomly selected to split or steal a jackpot of $1,000.

Beyond being of academic interest and entertaining, the game is used to model a variety of problems in sociology, biology, psychology, anthropology, business, law, politics, and nuclear warfare. Despite the fact that the prisoner's dilemma may be mathematically unresolvable, it is beneficial to study the models. Axelrod and Dion write, "Research along such lines will not only help unite theoretical and empirical work, even further than has so for been possible, but may also deepen our understanding of the initiation, maintenance, and further evolution of cooperation" [Axelrod and Dion 88].

In situations where all individuals benefit from collective action, there may be compelling reasons for an individual to opt out of the action yet still receive the benefit. The *free rider* dilemma is a multiplayer prisoner's dilemma deriving its name from public transit systems that do not check every passenger for a ticket. Spot checks are random, making it easy to board without a ticket if one so chooses. In some cases it's easy to hop the turnstile and ride for free. The rationalization is, "Why should I pay if I don't have to? I'll almost certainly not be caught and one less fare will not bankrupt the transit system." But, what if all riders behave this way? Then, the system would fail and all will suffer. Even if this were to happen, it is unlikely that an individual rider would feel guilty. As poet Stanislaw Lec describes it [Lec 02, p. 35], "No snowflake in an avalanche ever feels responsible."

In the US, public television requires viewer donations to survive. But, why subscribe voluntarily when you can watch for free? In fact, most viewers watch PBS without paying anything. But, if all viewers were to have this attitude, PBS would cease to exist. Any system where payment is based on an honor system is vulnerable to the free rider problem.

An interesting free rider dilemma is familiar to any citizen of a large society where there are elections to choose public officials and pass propositions. If voting is voluntary, a voter can make the case that it's not worth casting a vote. Isn't voting essential for a democracy to function? Doesn't every vote count in a fair election? Every vote does count but it is highly unlikely that any single voter will determine the election's outcome. Only in the event of a tie would one vote be decisive, and this will rarely, if ever, occur. Gelman, King, and Boscardin estimate this probability to be about 1 in 10 million for a close national election such as that which occurred in the US 1992 presidential election. The probability drops to less than 1 in 100 million in landslide elections such as the US 1972 presidential election [Gelman, King, and Boscardin 98]. In comparison, the risk of being murdered in the US on any given day is on the order of 1 in 7 million and the risk of being killed in a motor vehicle accident on any single day is about 1 in 3 million. One could make the argument that the cost of voting (driving to the polling place, just being out and about, or even the cost of postage for a mail ballot) exceeds the expected benefit (zero if your vote is not decisive) and rationally one shouldn't bother voting at all.

US citizens are encouraged to vote; many don't. Typically 50 to 60 percent of eligible voters cast votes in presidential elections and the rate drops to as low as 10 percent in many primaries. One option: compulsory voting. Australia, Austria, Belgium, and a host of other countries have forms of mandatory voting with penalties for not voting ranging from fines or community service to possible prison time. Beyond elimination of the free rider problem, compulsory voting ensures the government represents a majority of the population, not just those who are politically active or who represent special interest groups. Hopefully voters will be encouraged to research the candidates and issues, thus voting more intelligently, resulting in a more beneficial election outcome. Voters choosing not to support a choice have the option of marking "none of the above" on their ballot.

There is a down side to mandatory voting, and the vast majority of Americans oppose it. In the US, voting is largely thought of as a right, not an enforceable duty. Forcing citizens to vote takes away their freedom of choice. And, why should individuals who are ignorant of current candidates and issues be forced to make decisions about these matters? Shouldn't these choices be left to those who are informed and truly care?

A prisoner's dilemma scenario similar to the free rider problem is described by Garrett Hardin as *the tragedy of the commons* [Hardin 68]. Imagine a pasture open to all herdsman, each trying to keep as many cattle as possible on the common pasture. Herdsmen increase the size of their flocks and eventually the carrying capacity of the pasture will be reached. A herdsman believes he can easily add one more animal to the herd and benefit from its sale. The down side is minimal as the overgrazing caused by one additional animal is shared by all the herdsmen. His gain outweighs the minimal loss, and he is compelled to add the additional animal. But, each rational herdsman will reach the same conclusion and the common pasture will be ruined.

The problem exists wherever there is a limited environmental resource available for public consumption. Problems associated with water conservation, environmental pollution, and over fishing are all of this nature. All agree that cooperation is required, but voluntary compliance is unlikely to take place. Slogans like "Give a hoot, don't pollute!" and "Don't be a litterbug" are not enough to discourage littering. So, laws are passed and enforced; violators are fined or prosecuted.

Biomedical research and health-related clinical trials come with a unique triple set of prisoner's dilemma hindrances. The purpose of clinical trials is to answer questions regarding the effectiveness and safety of new treatments and drugs. In the US such trials follow a strict protocol designed to gather information and make the trials as safe as possible for the research subjects. But, at different stages of the process, a free rider situation emerges where individuals acting in their own best interest do so at the expense of the common good.

The first dilemma occurs when the research subjects are being recruited. Studies may require subjects with pre-existing illnesses and/or healthy individuals. But, ultimately it is up to each individual whether or not they choose to participate. There are pros and cons to be considered. The advantages include the following:

1. The subject may receive a promising new treatment along with expert medical care, which might be otherwise unavailable.
2. The treatment being evaluated may be less invasive and come with fewer side effects than conventional treatments.
3. There are altruistic considerations; participation in such a trial will help others with the same disease.
4. There are psychological benefits (which may translate to medical benefits) in playing an active role in one's own health care.
5. There may be some financial compensation including travel expenses.

Then there are the downsides:

1. The drug or treatment being evaluated may be less effective than conventional drugs and treatments.
2. The side effects of the drugs or treatment being evaluated may be worse than those of conventional treatments. The side effects may be unpleasant, if not serious and life-threatening.
3. If the trial is randomized, there may be no choice as to the type of treatment.
4. In a blinded study, the subject may be part of the control group, receiving a placebo having no known treatment value.
5. The trial may require a significant time investment. Clinical trials often require more visits to clinics and hospitals than conventional treatments.

6. Despite the possibility of financial compensation, all costs may not be covered and the subject's medical insurance may not cover the costs.
7. The treatment may prove to be completely ineffective.

For a significant number of prospective subjects, the cons outweigh the pros and the decision will be made not to participate. But, if we all felt this way, and many of us do, there would be no clinical trials. Safe and effective drugs and treatments would never be developed and we all would suffer. This is the classic free rider version of the prisoner's dilemma where individuals choosing not to participate in clinical trials get to _ride for free_, on the backs of those that do participate.

The second of the three dilemmas involves the possibility of pharmaceutical and biotech companies sharing information with regard to research, discovery, development, and clinical trials. Companies, medical providers, and patients would all benefit as such sharing expedites the research and development of drugs and treatments. But does it happen? Rarely. The "for profit" structure forces research and development companies to conceal such data from their competitors. Society would benefit from mutual sharing, but good business practice prevents this and the development is hindered. The companies would make the case, "Why give away what we have paid for? Let the other companies share their data, if they are so inclined."

In an article entitled "Can Science Be a Business? Lessons from Biotech," Gary Pisano makes the case that businesses engaged in basic sciences need a new design in order to allow for the integration of critical knowledge [Pisano 96]. The current state of affairs creates "islands of expertise" where individual firms lock up data and other information. This hinders the advance of knowledge with respect to drug research and development, health care, and other government-funded scientific research.

To completely rectify the problem, a national healthcare data sharing system could be put in place to facilitate the sale and/or sharing of such data between companies. Those against government expansion and national health would surely be opposed to such a concept. But, without sweeping changes, the players (research and development companies) are trapped in a prisoner's dilemma not easily escaped by the individual action of the firms.

The third dilemma involves the funding. Research and development are funded by pharmaceutical companies but may also be funded by physicians, medical institutions, foundations, voluntary groups, and federal agencies. A free rider dilemma develops when such groups are reluctant to sponsor such studies, relying on others to do so. The problem is particularly acute with respect to developing new antibiotics to combat antibiotic-resistant bacteria. Life-saving antibiotics may lose their effectiveness over time because bacteria can develop resistance to the drugs. A contributing factor is the mass-marketing of the antibiotics by drug companies, each trying to increase its own share of the market. (This leads to yet another prisoner's dilemma, in the form of a "tragedy of the commons.") Each year drug-resistant bacteria kill thousands of people in the US alone, and an epidemic could potentially harm millions. So, why is it that over the past 25 years public- and private-sector research and development of antibiotics has slowed dramatically?

1. There are existing generic antibiotics that effectively combat the large majority of infections.
2. The increased popularity of generics reduces the incentive to produce new medications.
3. Antibiotics may not be as profitable as other medications due to their relatively short duration regimes (five to ten days).
4. The rapid development of resistance for some bacteria decreases the long-term profitability of the antibiotic.

As the problem approaches crisis level, US federal agencies are suggesting incentives to spur the development of new antibiotics. These include financial incentives in the form of tax credits, patent extensions, and measures to reduce the cost of clinical trials.

It is now common for clinical trials to include an economic analysis of the cost-effectiveness of the drugs under consideration. Today's drugs can be highly expensive, and an early economic assessment is beneficial. But, who will fund this component of clinical trials? A free rider dilemma explains the reluctance of funding organizations to participate. Potential funding organizations independently decide whether or not to fund the study. If two or more agree to do so, the costs are evenly divided. If none agree then there will be no funding and the study may be cancelled. If only one organization agrees to fund the study, then it will bear the entire cost. The dominant strategy is to not fund whether or not others decide to do so.

As an example, a 1992 clinical trial was proposed by the Cancer and Leukemia Group B (CALGB), a group of academic and community hospitals in the United States. Its purpose was to compare the costs associated with two alternative treatment strategies. No funding agency was willing to support the financial component of the trial, and after two years of planning, the trial was canceled [Bennett et al. 95]. New systems of financial support are required to avoid such obstructions to healthcare research. These include the possibility of a central authority making the decision for all potential funding agencies or some form of tax where all organizations contribute money to fund such studies.

The Nash Arbitration Scheme

Certain non-zero-sum games (e.g., the pizza-pâté game, the prisoner's dilemma, chicken) have dilemmas that cease to exist if we allow players to enter into a pre-play negotiated agreement with respect to the payoff. Whether such a settlement is achieved by the players or an outside arbitrator, the settlement payoff should be both efficient and fair. By *efficient* we mean there would be no settlement that both players prefer to the given one. But what does *fair* mean? Can we expect such an agreement to be both efficient and fair? Would such a solution be unique?

Nobel Prize winner in economics, John Nash, gives an elegant arbitration scheme that, under intuitive assumptions, produces a unique solution.

Nash's method assumes payoffs are ordered pairs of utilities (r,c), where r and c represent the intrinsic worth of the outcome to the row and column players, respectively. So, if the row player prefers outcome 1 with payoff (r_1,c_1) to outcome 2 with payoff (r_2,c_2), it would follow that $r_1 > r_2$. A similar requirement holds for the column player. Also, if the row player is indifferent to payoff (r,c) and a lottery with probability p of payoff (r_1,c_1) and probability $1 - p$ of payoff (r_2,c_2), then $r = pr_1 + (1 - p)r_2$. Again, a similar requirement holds for the column player.

A player's utilities and the differences between pairs of utilities are to reflect the player's relative preferences. In doing so, the utility function for a player is not unique. We say two utility functions r and ρ for the row player are equivalent if there exists constants $a > 0$ and b such that $\rho = ar + b$. For example, let $r_1 = 1, r_2 = 2$, and $r_3 = 3$ denote the row players preferences for outcomes 1, 2, and 3, respectively. Using $a = 2$ and $b = -1$, we arrive at the equivalent set of utilities $\rho_1 = 1, \rho_2 = 3$, and $\rho_3 = 5$. They are equivalent in the sense that both reflect the order and relative strength of the row player's preferences. Order preservation, in this case, is obvious. Also, note that the r values indicate that the row player is indifferent to outcome $2(r_2 = 2)$ and a lottery having equiprobable outcomes $1(r_1 = 1)$ and $3(r_3 = 3)$. The ρ values show the same. There are infinitely many sets of equivalent utilities that preserve the row player's preferences. All are of the form $ar + b$ for constants $a > 0$ and b. Transformations of utility functions by the formula $\rho = ar + b$ possess a quality known as *linear invariance*. The term *linear* is a bit technical and its definition need not concern

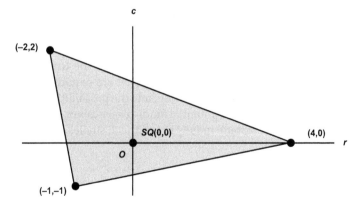

Figure 8.2. Arbitration polygon.

us here. The term *invariance* indicates that the player's relative preferences do not vary and are preserved under such transformations. In similar fashion, equivalent utility functions exist for the column player.

Our friends Ron and Carla employ an arbitrator to determine an efficient and fair payoff, $E^* = (r^*, c^*)$, associated with a set of possible payoff points, each of the form (r,c). If they fail to come to an agreement, they will receive a *status quo* payoff, represented by point SQ. Our discussion is simplified by assuming that SQ corresponds to $(0,0)$. We may do so because if SQ were located elsewhere, we could apply a linear transformation of the form previously described that would move SQ to $(0,0)$ and preserve both players' relative preferences. Figure 8.2 gives the arbitration (payoff) polygon for the conflict to be negotiated. The polygon is formed from the set of four payoff points, one of which is the status quo point SQ. The shaded region represents all possible payoffs, including expected payoffs that could be obtained by lottery involving the initial possible payoff points.

At this point the reader may ask, "Why not simply agree on the payoff (4,0), which maximizes $r + c$? This is the most Ron and Carla could receive together, and they could fairly split the difference. But, this would be meaningless as the utility functions for each are not unique. Utilities only reflect intrapersonal preferences and can't be added to the utilities of another player in a meaningful way.

Nash gives conditions that any efficient and fair solution should satisfy:

Condition 1

The solution point $E^* = (r^*, c^*)$ must be such that $r^* \geq 0$ and $c^* \geq 0$. That is, neither the row player nor the column player should accept a negotiated payoff less than what they would receive if negotiations were to fail or never occur.

Condition 2

There should be no point (r,c) in the arbitration polygon such that $r > r^*$ and $c \geq c^*$ or $r \geq r^*$ and $c > c^*$. This guarantees the efficiency of E^*.

Condition 3

If the row player's utilities are multiplied by the positive constant a, and the column player's utilities multiplied by the positive constant b, the solution point will be transformed to (ar^*, bc^*). This is a consequence of linear invariance.

Condition 4

If the payoff polygon is symmetric about the line $r = c$, then E^* should be on this line. This is an intuitive component of fairness, eliminating discrimination.

Condition 5

Suppose that E^* is the solution (fair and efficient) for arbitration polygon Q. Let P be an arbitration polygon that is completely contained in Q that contains both $(0,0)$ and E^*. Then E^* should also be the solution for P. This is commonly referred to as the *independence of irrelevant alternatives* (IIA) condition.

The first four conditions seem reasonable and are generally accepted. Condition 5 is controversial and some refuse to accept it. But consider this: You are at a friend's house for dinner, and your host offers you a beer. He asks if you prefer Amstel, Budweiser, or Corona. Your favorite of the three is Amstel. But, just before you make your request, your host tells you that Corona is no longer available. Would you still prefer Amstel, knowing that Corona is no longer available? Of course you would! Amstel is your preference and remains so, despite the elimination of Corona as an irrelevant alternative.

Nash's bargaining scheme asserts that there is one and only one point within the arbitration polygon that satisfies all five conditions [Nash 50] and therefore it is this unique point that serves as E^*. Three cases are considered:

Case 1

If $r \leq 0$ and $c \leq 0$ throughout the payoff polygon, $E^* = SQ = (0,0)$.

Case 2

If $r = 0$ and $c > 0$ throughout the payoff polygon, $E^* = (0, c_{max})$. Similarly, if $r > 0$ and $c = 0$ throughout the payoff polygon, $E^* = (r_{max}, 0)$.

Case 3

If a region of the payoff polygon contains points where $r > 0$ and $c > 0$ (as in Figure 8.2), E^* is the point where the product rc is maximized.

It's easy to see that, for the first two cases, the solution is as given and that all five conditions are satisfied. Of interest is case 3. Assume the product rc is maximized at the point $N = (r_N, c_N)$, known as the Nash point. We wish to show that $E^* = N$. It's reasonably clear that the payoff N satisfies all five conditions. We need only show that N is unique; that is, no other point (case 3) satisfies the five conditions. The simple proof follows. Maximizing the product does not, in and of itself, produce an efficient and fair solution. We will show, however, that the Nash point is the only point that satisfies Nash's five conditions. It must therefore be that $E^* = N$.

For any given payoff polygon having $SQ = (0,0)$ and containing points such that $r > 0$ and $c > 0$, change the utility scales for the players so that N is trans-

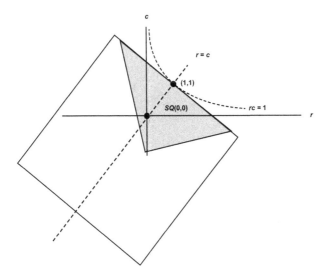

Figure 8.3. Arbitration polygon imbedded within a larger square.

formed to (1,1). This is done by multiplying all possible values of r by $1/r_N$ and all possible values of c by $1/c_N$. These two linear invariant transformations preserve the relative intrapersonal preferences of the players. The point (1,1) is at the vertex of the upper right branch of the hyperbola $rc = 1$, as shown in Figure 8.3. Since N maximizes rc, points in the transformed arbitration polygon lies entirely on or below the hyperbola, touching it at (1,1). Since the arbitration polygon is convex, it must lie on or below the tangent line to the hyperbola at (1,1). The equation of this line is $r + c = 2$.

Now enclose the arbitration polygon in a large square with one side on this tangent line so that it is symmetric about the line $r = c$ as shown in Figure 8.3.

By conditions 1, 2, and 4, (1,1) is the only point satisfying the five Nash conditions for the large square region and must therefore be the solution for the large square region. Condition 5 requires that (1,1) also be the solution point for the transformed arbitration polygon (shaded). Condition 3 implies that N must be the solution for the original arbitration polygon and we are done. The efficient and fair solution is $E^* = N$ because no other expected payoff satisfies the five Nash conditions.

In the following appendix, the Nash scheme is applied to the arbitration problem given at the beginning of this section and depicted in Figure 8.2.

Appendix

Applying this result to the arbitration problem given in Figure 8.2, rc is maximized on the upper side of the triangular boundary at the point (2,2/3). To see why, note that the equation of the upper leg of the triangle is $c = \frac{4}{3} - \frac{r}{3}$, for $-2 \leq r \leq 4$. To maximize rc (here we need calculus again), differentiate with respect to r and set the derivative equal to 0:

$$\frac{d}{dr}(rc) = 0 \implies \frac{d}{dr}\left(\frac{4r}{3} - \frac{r^2}{3}\right) = 0 \implies \frac{4}{3} - \frac{2r}{3} = 0 \implies r = 2, c = \frac{2}{3}.$$

So, $E^* = N = (2, \frac{2}{3})$.

We need to find probabilities p and $1 - p$ to define the lottery that generates this expected payoff. In terms of Ron's payoff,

$$p(-2) + (1-p)(4) = 2 \implies p = \frac{1}{3}, 1 - p = \frac{2}{3}.$$

So, the solution payoff $E^* = N$ is achieved by lottery, where $(-2,2)$ occurs with probability $1/3$ and $(4,0)$ occurs with probability $2/3$.

Newcomb's Paradox

We must believe in free will; we have no choice.
—Isaac Bashevis

I s it your free choice to read the next few paragraphs of this chapter? Or, are there forces requiring you do so? The centuries-old free will versus determinism debate may be impossible to resolve as it requires one to consider one's own thoughts and actions. It's a nasty self-referential problem debated by philosophers and theologians. You may strongly believe in your own free will. But, conceivably this very belief and all other thoughts and actions are forced, governed in mysterious, incomprehensible ways. If the mannequin in the store window believes it to be her choice to remain standing, how could you convince her otherwise?

Newcomb's paradox is a simply stated envelope problem, much like those of Chapter 5. Maximizing expectation offers one solution; but, it flatly contradicts a belief in free will. Read the problem and decide which of the two options works best for you. Your decision (if we agree that you may freely make one) will draw on your innermost feelings regarding free will and determinism.

Two closed boxes are placed before you, as in Figure 9.1. The box on the left (B1) is known to contain $1,000. The box on the right (B2) contains either $1 million or $0. You, the player, know nothing more about the contents of B2, only that it contains $1 million or $0. You are given two options—take what is in both boxes or take only what is in B2. The contents of B2 is determined beforehand by an omniscient being having the ability to predict future actions with a high degree of reliability. If the being predicts you will take the contents of both boxes, it places nothing in B2. If the being predicts you will take only what is in B2, it places $1 million in B2. (If the being predicts you will toss a coin, making your choice random, it

$1,000 $0 or $1 million

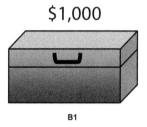   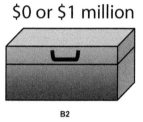

B1   B2

Figure 9.1. Newcomb's paradox.

places nothing in B2.) Either way, B1 contains $1,000. The boxes are now closed and you make your selection. Do you choose both boxes or B2 alone?

If you believe the answer is obvious you're forgetting this book's title.

Dominance vs. Expectation

The paradox arises because solidly rational arguments can be given supporting each of the two options. The paradox is presented here in dialogue form. At first glance it may seem obvious that you should choose both boxes. You are applying a dominance argument that, regardless of B2's contents, two boxes are better than one. In the dialogue below, the player makes the case for choosing both boxes and is challenged by a devil's advocate (DA), making the case for choosing B2 alone.

> PLAYER: The choice is obvious. Two boxes are better than one, regardless of their contents. So, I'll take both B1 and B2.
>
> DA: Would you be a bit more precise?
>
> PLAYER: How much more precise can I be? We know the contents of B1 is $1,000. The contents of B2 is unknown; let's call it x. If I choose both boxes, I get $1,000 + x. If I choose B2 alone I get x. No matter how you look at it, $1,000 + $x > x and I'm better off with both boxes.
>
> DA: May I make the case for choosing B2 alone?
>
> PLAYER: Give it your best shot!
>
> DA: Let's not forget about the being and its prediction. You understand we assume the being has an excellent track record when it comes to predictions of this sort. It has studied human behavior and knows you quite well. We must assume its prediction will almost certainly be correct.
>
> PLAYER: I understand.
>
> DA: If you do as you say and select both boxes, will this act have been predicted by the being?
>
> PLAYER: Almost certainly.
>
> DA: And the being will have placed nothing in B2. Right?
>
> PLAYER: Yes.
>
> DA: So, by choosing both B1 and B2, you will receive a total of $1,000.
>
> PLAYER: True.

DA: But consider choosing B2 alone. Assuming you do so, will the being have predicted so?
PLAYER: Yes.
DA: Under these conditions the being will have placed $1 million in B2 and this will be yours.
PLAYER: I see where you're going with this.
DA: I knew you would see it my way. If you prefer $1 million to $1,000, and who wouldn't, you should select only B2.
PLAYER: You're correct. I'll choose only B2.

After giving the problem some thought, you might now favor choosing only B2, based on maximizing the expected value of your choice as described above. The player in the dialogue below prefers this option and is again challenged by a devil's advocate who makes the case for choosing both boxes.

PLAYER: I choose B2. This will almost certainly have been predicted and I expect to collect $1 million. May I open the box now and take my money?
DA: If you wish. But may I first make a few observations?
PLAYER: Now what?
DA: B2 is yours, one way or the other. Don't open it just yet; but, you will be opening it soon, no matter which of the two options you select. So it's yours, either way.
PLAYER: Good. I agree.
DA: Now, would you like an additional $1,000 with no strings attached?
PLAYER: No strings attached? Sure.
DA: Then take B1 also. It contains $1,000 and it's yours for the asking.
PLAYER: But, if I take both boxes, there will be nothing in B2.
DA: But you just said it contains $1 million! Will choosing B1 cause the contents of B2 to disappear? Of course not! The contents of B2 is fixed and there is no backward causality. The being has done its thing and placed either nothing or $1 million in B2. As far as you're concerned, the being no longer exists. Nothing you do now will change the being's prediction or the contents of B2. I'm simply asking if you want an additional $1,000 to whatever you are now guaranteed to receive when you open B2.
PLAYER: Uh, I'll need some time to think about this.

Two days later:

DA: Have I convinced you to take both boxes?
PLAYER: I understand your argument but I'm still going to choose only B2.
DA: Fine. May I bring up one last point?
PLAYER: Make it quick. I want my money.
DA: While you were gone I peaked inside B2 and I know its contents.
PLAYER: You can't do that! It's cheating.
DA: It's not cheating if *I* do it. You're the player, not I. And, I'm not going to tell you what I saw. I'm simply letting you know that *I know* whether or not B2 contains the $1 million.

PLAYER: Why are you telling me this?
DA: True to my name, I'm trying to get you to switch positions. When I tell you I know the contents of B2, I'm making the point that its contents are now fixed and can't be changed. So, be that as it is, do you want to take B1 as well and collect the additional $1,000?
PLAYER: You're good. Where did you go to law school? I'll take both boxes.

The problem was conceived in 1960 by William Newcomb, a professor and theoretical physicist at the University of California's Lawrence Livermore Laboratory. It received no significant attention until Harvard philosophy professor Robert Nozick published "Newcomb's Problem and Two Principles of Choice" [Nozick 69]. Martin Gardner presented the problem in his July 1973 "Mathematical Games" column of *Scientific American* [Gardner 73]. Gardner's column ran from 1956 to 1971, and he reports receiving more mail in response to Newcomb's problem than any other column he wrote for the magazine [Gardner 96, p. 412]. Nozick's reaction to the mail appears in the March 1974 edition of the magazine [Gardner and Nozick 74]. Both articles can be found in Gardner's book, *Knotted Doughnuts and Other Mathematical Entertainments* [Gardner 86, pp. 155–175].

(The last three chapters of this book are connected in a curious way. William Newcomb discovered Newcomb's paradox while thinking about the prisoner's dilemma game presented in Chapter 8. Prisoner's dilemma and its connection to Newcomb's paradox are discussed at the end of this chapter. Benford's law, the subject of the last chapter of this book, was first discovered by the famous astronomer Simon Newcomb, the great-great-uncle of William Newcomb.)

There are principles solidly supporting each of the opposing options. The dominance principle, as discussed in Chapter 3 and depicted in Matrix 9.1, provides a solid argument for choosing both boxes.

Regardless of the being's prediction, you gain more by choosing both boxes. Clearly row 2 dominates row 1, and the dominance principle compels the player to take both boxes.

The other option, choosing only B2, can be justified using the principle of maximizing expectation. Rather than thinking in terms of the being's two possi-

	Being	
	Predicts you take only B2	Predicts you take both B1 and B2
You Take only B2	$1,000,000	$0
Take both B1 and B2	$1,001,000	$1,000

Matrix 9.1. Dominance principle and Newcomb's paradox.

	Being	
	Predicts correctly (p)	Predicts incorrectly (1–p)
You Take only B2	$1,000,000	$0
Take both B1 and B2	$1,000	$1,001,000

Matrix 9.2. Expected value principle and Newcomb's paradox.

ble predictions, we consider two states of nature that occur following the prediction. Ultimately, the being's prediction is either correct or incorrect, as shown in Matrix 9.2. Assume the being predicts correctly with probability p and therefore incorrectly with probability $1 - p$.

Your expectation in choosing B2 alone is given by

$$E_{B2} = p(\$1,000,000) + (1 - p)(\$0) = \$1,000,000p,$$

and your expectation in choosing both boxes is

$$E_{B1\&B2} = p(\$1,000) + (1 - p)(\$1,001,000) = \$1,001,000 - \$1,000,000p.$$

If $p > \frac{1001}{2000} = .5005$, then $E_{B2} > E_{B1\&B2}$ and you are better off choosing B2 alone.

The argument holds if the being's probability of success is only slightly greater than 50 percent. In principle, such a talent can't be ruled out. If we assume the being has studied human nature and the player's background (education, personality, etc.), it is quite conceivable the being predicts correctly most of the time.

Many explanations and "resolutions" of Newcomb's paradox have been put forward since Nozick first published the problem in 1969. Most sidestep the problem by saying, in essence, "If we look at things this way, the paradox disappears." This may be true; but, to completely resolve the issue, we must show what is logically incorrect about looking at it "the other way." Nozick writes [Nozick 69, p. 117],

> Given two such compelling opposing arguments, it will not do to rest content with one's belief that one knows what to do. Nor will it do to just repeat one of the arguments loudly and slowly. One must also disarm the opposing argument; explain away its force while showing its due respect.

At some level, the problem is tied to the free will versus determinism debate. Are you, the player, capable of freely choosing an option if the being has already made its prediction and we assume such predictions are highly accurate?

If you believe in free will, then you might choose both boxes, hoping the being has erred in predicting you will choose only B2. In such a case, you get the maximum possible payoff of $1,001,000. But, your choice may only feel free, much like the mannequin in the store window "believing" it is her choice to remain standing in place. On the other hand, if you believe your fate is set as predicted by the being, you will comfortably accept $1 million by "choosing" B2 alone.

> As author of this book, I must frankly admit that I've thought about this paradox for many years and I'm unable to resolve it. There simply may be no resolution, a fact that elevates the problem far and above those mathematical exercises with "answers to be found at the back of the book." It's not the only paradox in mathematics and I'll try not to lose any more sleep over this one. Like the bumper sticker says, "I feel better now that I've given up hope." I have, however, answered the other question associated with this problem. Would I choose B2 alone or both boxes? This one is easy for me to answer, as long as I'm not required to defend it. I choose B2 alone and I'll enjoy spending my $1 million. Don't ask how I refute the dominance argument. I don't. Must I? I'm reminded of John von Neumann's comment given in this book's introduction. "In mathematics you don't understand things. You just get used to them." I'll have no problem getting used to $1 million. Does this make me a determinist? Maybe not. I still plan on paying my VISA bill this month and I wake up early each morning to work on this manuscript. A true determinist would not do such things. But, why can't I get the words to *Que Sera, Sera* out of my head?

Thinking "outside the box(es)," Gardner asks [Gardner 73, pp. 108–109],

> Can it be that Newcomb's paradox validates free will by invalidating the possibility, in principle, of a predictor capable of guessing a person's choice between two equally rational actions with better than 50 percent accuracy?

The decision is yours. Maybe.

Newcomb + Newcomb = Prisoner's Dilemma

Some choose to dismiss discussions of free will, determinism, and predictability as nothing more than fanciful recreation, in which case it would be pointless searching for real-world manifestations of Newcomb's paradox. But, if one believes in predestination, real-world Newcomb phenomena emerge. Michael Resnik gives two examples [Resnik 85, p. 111].

Calvinism is a Christian belief system, first promoted by the French church reformer John Calvin (1509–1564). One of the tenets of Calvinism asserts that all humans, independent of their actions, are preordained by God to eternal salvation or eternal damnation. God made the election before the universe was created and nothing one does during their lifetime can change this. A Calvinist, tempted to sin, faces a dilemma similar to that of the player who must choose between taking what is in both boxes or taking only what is in B2. The dominance argument suggests the Calvinist may as well go ahead and sin, assuming there is pleasure associated with sinning. Doing so will not effect the likelihood of eternal salvation, as this has been predetermined by God. The expected value argument warns the Calvinist not to sin, as doing so would be a sure sign of one's predestined eternal damnation.

Predestination, assuming it exists, may have a biological foundation in the form of a genetic predisposition for type A personality characteristics such as competitiveness, impatience, and hostility, known to be associated with heart disease. It's generally agreed that smoking is a direct cause of heart disease. But, assume there exists some genetic defect that predisposes an individual to having the type A personality; and, it is this defect that is the cause of heart disease, not smoking. Further assume that other diseases associated with smoking, such as lung cancer and emphysema, are caused by this genetic defect and not by smoking itself. (Correlation does not imply causation!) If one finds smoking pleasurable and must decide whether or not to smoke, one is in a dilemma similar to that of the Calvinist and the box-choosing player. On the one hand, why not smoke? Assuming smoking is not the cause of these diseases, then choosing to smoke will not increase the likelihood of the disease. This likelihood is predetermined by genetic factors. On the other hand, smokers are more likely to develop heart disease. So, why would one knowingly place themselves at risk?

Curiously, William Newcomb discovered this paradox while pondering the prisoner's dilemma game discussed in the preceding chapter. The problems are linked, if not one and the same! Newcomb's paradox is formulated as a solitaire game. A single player chooses one of two options so as to maximize a monetary gain. The omniscient being is an actor, not a player; it neither wins nor loses when it makes its prediction. But, envision a Newcomb game of two identically rational and omniscient players (Being 1 and Being 2), each predicting what the other (and hence themselves) would do when faced with the choices of Newcomb's paradox. Consider predictions and choices being made simultaneously and combine the two payoff matrices into the single matrix as shown in Matrix 9.3.

		Being 2	
		Take B2 only	Take both B1 and B2
Being 1	Take B2 only	($1,000,000, $1,000,000)	($0, $1,001,000)
	Take both B1 and B2	($1,001,000, $0)	($1,000, $1,000)

Matrix 9.3. Newcomb's paradox as a prisoner's dilemma.

Look familiar? Side-by-side Newcomb choices form the classic prisoner's dilemma game! A full discussion of the relationship between Newcomb's paradox and the prisoner's dilemma is given in [Brams 83, pp. 52–54] and [Lewis 79].

Real-world instances of Newcomb's paradox should not be ruled out. Lewis writes [Lewis 79, p. 240],

> Some have fended off the lessons of Newcomb's problem by saying: "Let us not have, or let us not rely on, any intuitions about what is rational in goofball cases so unlike the decision problems of real life." But prisoners' dilemmas are deplorably common in real life. They are the most down-to-earth versions of Newcomb's problem now available.

Benford's Law

There's something happening here.
What it is ain't exactly clear.
                                                        —Stephen Stills

B enford's law had its beginnings over a century ago when American astronomer Simon Newcomb noticed something curious about tables of common
logarithms. Pages at the beginning of the table appeared significantly more worn
than pages near the end, suggesting mathematicians and scientists were looking
up numbers beginning with lower digits (1, 2, ...) more often than higher digits
(..., 8, 9). (Simon Newcomb is the great-great-uncle of William Newcomb, the
creator of Newcomb's paradox). This would suggest that numbers being looked
up (measures of naturally occurring physical quantities) begin more often with 1
than any other single digit, a highly unexpected phenomenon.

Interest in Newcomb's discovery dropped off until the law was rediscov-
ered in 1938 by Frank Benford, a physicist working for General Electric. Empirical evidence supports Newcomb's discovery, now known as Benford's law; but,
mathematical validation was slow in coming. A mathematical law is proven as a
theorem by a sequence of statements following a given set of axioms. The proof
is eternal and stands the test of time, unless the proof itself contains an error. On
the other hand, physical law is evidence based and may be updated or overturned
with additional discoveries. Benford's law is a strange mix of both physical (real-
world data) and mathematical law. Not all data sets comply, and this has hindered
its understanding and acceptance. Today we have a clearer understanding of the
phenomenon. Applications include computer design, testing of mathematical
models, and detection of accounting fraud.

Simon Newcomb's Discovery

Logarithms are used in a wide variety of scientific calculations including the chemical measure of acidity (pH), Richter scale magnitudes of earthquakes, and sound intensity measured in decibels (dB). The common (base-10) logarithm of a positive real number is the power to which 10 must be raised to yield that number. That is, $y = \log x$ if and only if $x = 10^y$. For example, $\log 10 = 1$ because $10 = 10^1$, $\log 100 = 2$ because $100 = 10^2$, $\log 1000 = 3$ because $1000 = 10^3$, etc. Finding the logarithm of a real number that is not a perfect power of 10 requires a computer or calculator to approximate the logarithm. Prior to digital computing, tables of logarithms were used for the approximations in much the same way as square-root tables were used to approximate square roots. The process of finding the common log of a number using a log table begins by writing the number in scientific notation—in the form $a \times 10^n$, where $1 \le a < 10$ and n is an integer. (Note that $\log(a \times 10^n) = \log a + \log 10^n = \log a + n$ or $n + \log a$.) For example, to find the logarithm of 1,230 we first write this number as 1.23×10^3. The characteristic, or integral part, of the logarithm is n, or in this case 3. The mantissa, or decimal part, of the logarithm is found to be .0899 by locating a, in this case 1.23, in the table and reading off the four-place mantissa. So, $\log 1230 \approx 3 + .0899 = 3.0899$. Of importance here is the fact that it's a number's leading digit(s) that determine its location in a table.

In 1881 American astronomer Simon Newcomb noticed that tables of logarithms were worn heavily at their beginning and less worn toward the end of the table, indicating that numbers beginning with the low digits 1,2, ... were looked up more often than numbers beginning with the higher digits ..., 8,9. It would

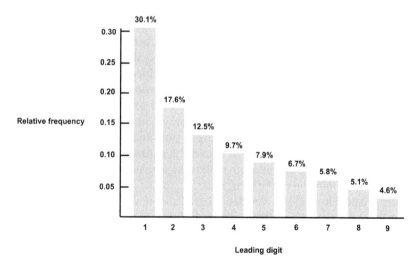

Figure 10.1. Simon Newcomb's discovery.

follow that numbers with low first digits would occur in nature more frequently than those beginning with high digits. Naively, one might expect the first digits of naturally occurring numbers to occur uniformly, each with relative frequency $1/9 \approx 11$ percent. (Zero is not considered a leading digit.) Further investigation led Newcomb to believe leading digits actually occur with frequencies as in Figure 10.1.

What could possibly explain this unexpectedly skewed right distribution, where numbers beginning with 1 are more than six times as likely to occur than numbers beginning with 9? It's clear that many data sets do not conform. Healthy adult body temperatures measured in degrees Fahrenheit would fail to comply as all data would begin with the digit 9. Random lottery numbers would also fail to comply as their distribution is uniform over a restricted domain. So, why should any set of data satisfy the law? The law does, in fact, hold for many different types of data sources including population sizes, street addresses, land area of nations, baseball statistics, and physical constants. It holds extremely well if the data is sampled from a multitude of sources, as reflected by the worn first pages of logarithm tables. Benford suggests the law applies to *anomalous numbers*, numbers that are not confined to restricted domains and are allowed to vary over several orders of magnitude.

Benford's Law

Both Newcomb and Benford arrived at the same conclusion. The probability (or theoretical relative frequency) of a number starting with the digit d is $\log(1+\frac{1}{d})$, where "log" denotes the common, base-10 logarithm. So, the probability of the first digit being 1 is $\log(1+\frac{1}{1}) = \log 2 \approx .301$, and the probability of it being 9 is $\log(1+\frac{1}{9}) \approx .046$. The law can be generalized to apply to a string of starting digits. For example, the probability, or relative frequency, of a number starting with the digits 123 is given by $\log(1+\frac{1}{123}) \approx .0035$. The formula can be used to find probabilities for digits in any position. For example, the probability of the second position digit being a 1 is found by summing the probabilities of the number's two leading digits being 11, 21, 31, ..., 91. In general, the probability of the second position digit being equal to d is given by

$$\sum_{k=1}^{9} \log\left[1 + (10k+d)^{-1}\right], d = 0, 1, 2, ..., 9.$$

As is the case with the first position digit, second position digit probabilities decrease as d increases. The approximate second position digit probabilities are as follows: 0 (12.0%), 1 (11.4%), 2 (10.9%), 3 (10.4%), 4 (10.0%), 5 (9.7%), 6 (9.3%), 7 (9.0%), 8 (8.8%), and 9 (8.5%). The decreasing probabilities are similar to those for the first digit, though flatter and less pronounced. Digit frequencies for subsequent positions (3rd, 4th, ...) conform to similar but less biased distributions.

First Digit

Data Type	1	2	3	4	5	6	7	8	9
Constants	36%	20%	8%	9%	8%	7%	2%	3%	7%
New York Times	31%	26%	11%	4%	8%	5%	10%	3%	2%
CA cities	29%	19%	13%	7%	7%	5%	11%	2%	7%
n^3	24%	16%	12%	10%	10%	8%	7%	7%	6%
$\tan n$	29%	17%	14%	10%	4%	11%	3%	8%	4%
Average (rounded)	30%	20%	12%	8%	7%	7%	7%	5%	5%

Table 10.1. Data conforming to Benford's law.

Table 10.1 gives leading digit frequencies for five diverse data samples, each of size 100. The first represents the first 100 entries in a table of fundamental physical constants. The second gives leading digits of 100 randomly chosen numbers from the first page of the New York Times (Business Section, January 7, 2011). The third row gives leading digit frequencies of populations for the first 100 alphabetically listed California cities given in a popular world atlas. The fourth row gives the leading digit frequencies of the first 100 values of n^3, $n = 1, 2, ..., 100$. And, the last row gives the leading digit frequencies of the first 100 values of the trigonometric function $\tan n$ (with n measured in radians).

Even the 81 products of the standard 9 × 9 multiplication table have a Benford-like distribution of leading digits: 1 (22.2%), 2 (18.5%), 3 (13.6%), 4 (14.8%), 5 (7.4%), 6 (8.6%), 7 (4.9%), 8 (6.2%), and 9 (3.7%).

The explanations for the unexpected behavior are as diverse as the data sources themselves. Benford's explanation begins with the assumption that real data is generated exponentially (or geometrically), more so than arithmetically, in which case successive data measurements would have a constant ratio, as opposed to a constant difference. The terms 1, 2, 4, 8, ... form a simple geometric sequence with a common ratio of 2. It's easy to see why such sequences have first digits as prescribed by Benford's law. As an example, begin with 1 and increase each term by 3 percent. (The common ratio is 1.03.) The sequence begins 1, 1.03, 1.0609, ... The first 24 terms begin with a 1, the next 14 terms begin with a 2 and the next nine terms begin with a 3. As the first digits increase, their frequency decreases. At the 79th term the first digit once again is a 1 and the cycle starts over. For the first 78 terms the first digit frequencies are 1 (24/78 ≈ 30.8%), 2 (14/78 ≈ 17.9%), 3 (9/78 ≈ 11.5%), 4 (8/78 ≈ 10.3%), 5 (6/78 ≈ 7.7%), 6 (5/78 ≈ 6.4%), 7 (5/78 ≈ 6.4%), 8 (4/78 ≈ 5.1%), and 9 (3/78 ≈ 3.8%), conforming reasonably well to the Benford distribution. If we assume continuous exponential growth, the first digit frequencies are precisely those prescribed by Benford. The derivation is given in this chapter's appendix.

Benford's reasoning relies on nature behaving exponentially. Having no way of knowing this with certainty, the above argument is not a *proof*.

Roger Pinkham of Rutgers University gives a unique explanation of the law by assuming *scale invariance* and avoiding Benford's assumption of exponential data. A distribution of digit frequencies is scale invariant if it remains the same regardless of the units used. For example, if a digit frequency distribution holds for river depths measured in feet, and if the distribution is scale invariant, the same distribution would exist if the depths were measured in yards, meters, fathoms, or any other unit of linear measurement. In 1961, Pinkham proved that Benford's law is scale invariant and, more importantly, it is the only first-digit law that is scale invariant [Pinkham 61]. (An informal proof is given in this chapter's appendix.) So, if there is some universal distribution law governing first digits, its universality would require it not depend on the units used, in which case Benford's law would apply.

To illustrate, consider the following length measurements, in yards: 15, 25, 35, 45, 55, 65, 75, 85, and 95. The first digit frequencies are uniformly distributed. Converting to feet (multiplying by 3), the data becomes 45, 75, 105, 135, 165, 195, 225, 255, and 285. The distribution of first digits has significantly changed, favoring the lower digits. This suggests uniform distributions are not scale invariant. Now consider a different set of measurements, also in yards: 10, 12, 14, 16, 18, 19, 20, 23, 26, 33, 37, 43, 47, 55, 65, 75, 85, and 95. The distribution is Benford-like, favoring the lower digits. Converting each measurement to feet gives respective measurements of 30, 36, 42, 48, 54, 57, 60, 69, 78, 99, 117, 129, 141, 165, 195, 225, 255, and 285. The distribution of first digits is roughly the same, still favoring the lower digits and indicating scale invariance.

Pinkham's scale invariance argument does not prove Benford's law because it is based on the assumption of there being a universal law governing first digits. In that not all sources of data satisfy Benford's law, the universal law assumption is questionable. B. J. Flehinger of IBM avoids this problem, offering an explanation of Benford's law without such assumptions [Flehinger 66]. She considers all significant figures of all numbers appearing in tables of physical constants, ignoring the sign and placement of the decimal point. Under these conditions we have a subset of the positive integers. What fraction (relative frequency, probability, density) of all positive integers begin with a 1? If our data set consists only of the first positive integer, {1}, then the answer is trivially 100 percent. As the data set expands to {1,2} then {1,2,3} up to {1, 2, 3, ..., 9}, the fraction drops to 1/2 then 1/3 and so on, finally reaching 1/9. When 10 is added, the fraction jumps up to 2/10 = 1/5 and continues to increase as the next nine positive integers are included. For the first 19 positive integers, the fraction that begin with 1 is 11/19; when we consider the first 99 positive integers, the fraction drops down again to 11/99.

Figure 10.2 plots this density for positive integers 1, 2, ..., *n*. (The horizontal axis is compressed.)

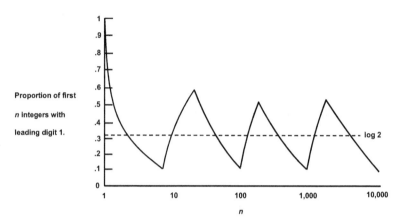

Figure 10.2. Probability that the first digit of n is 1.

As n increases, the fraction in question cycles between lows of approximately .1 and highs near .5. It fails to level off, or stabilize, and therefore the density in question fails to exist. But, in a sense, Benford's law is holding. Flehinger shows that the average height of the graph (an average of averages) for all possible values of n is log 2, conforming precisely to Benford's law. The law is confirmed for the other eight possible leading digits in similar fashion.

Using Flehinger's analysis, it's easily seen why street addresses conform to Benford's law. For any given street, the probability of a single home having an address beginning with 1 would fall roughly between .1 and .5 (ignoring the possibility of an address being 0 or 1). But, if we average the averages by considering all homes on all streets, the probability of a single home beginning with 1 is exactly log 2.

Does Flehinger's argument prove Benford's law? Hardly. For one, numbers found in tables are a subset of the positive integers and may not have the same first digit properties as those of all positive integers. Also, methods of averaging other than the one used by Flehinger do not give the value log 2.

Theodore Hill of the Georgia Institute of Technology offers what is generally regarded as the most rigorous explanation of the Benford phenomenon [Hill 95]. Benford's law applies to some data sources (street addresses, population size, physical constants), yet not to others (lottery numbers, telephone numbers, square-root tables). This is to be expected as different data sources have different distributions. We've already seen that uniform distributions fail to satisfy the law, whereas exponentially distributed data does. Hill shows that if distributions are selected at random, and then random samples are taken from each of the chosen distributions, then the combined sample data will converge to Benford's law. What works is a grand distribution, a *distribution of distributions*, something like Flehinger's *average of averages*. This clearly explains the worn first pages of logarithm tables, where numbers are looked up from a variety of sources. Numbers randomly selected from the pages of a newspaper will also comply as they come from a variety of sources (stock prices, weather data, sports statistics, etc.).

Hill's work is technical and not presented here. An article by R. M. Fewster appearing in *The American Statistician* gives a simpler explanation as to why the law applies [Fewster 09]. Any positive x can be written in scientific notation as $a \times 10^n$, where $1 \le a < 10$ and n is an integer. For any given distribution of x, consider the associated distribution of $\log x$ ($\log x = n + \log a$). The first digit of x is d whenever $d \le a < d + 1$, which holds if and only if $\log d \le \log a < \log (d + 1)$. Figure 10.3 represents the *probability density function* for $\log x$. The area under the curve between two values of $\log x$ on the horizontal baseline directly corresponds to the probability of $\log x$ being between these two values. Note that values of x with a first digit of 1 correspond to the vertical stripes, each of width

$$\log (1 + 1) - \log 1 = \log 2.$$

Consecutive integers n mark the left endpoint of each stripe. The probability of x beginning with the digit 1 is equal to the total area of all vertical stripes. Since the width of each stripe is $\log 2$ and there is one stripe per unit interval on the horizontal baseline, the stripes cover a proportion of the horizontal baseline equal to $\log 2$ and this crudely approximates the desired area, as Figure 10.3 should suggest. (The total area under the graph of a probability density function is always equal to 1.) The approximation is best for data sets having a large numbers of stripes, corresponding to data ranging over many orders of magnitude.

In general, x beginning with a first digit of d corresponds to equally spaced vertical stripes of width

$$\log(d+1) - \log d = \log\left(\frac{d+1}{d}\right) = \log\left(1+\frac{1}{d}\right).$$

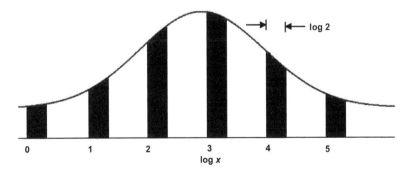

Figure 10.3. Fewster's stripes for $d = 1$.

Intuitively, if there were more stripes, and if the curve remained sufficiently smooth, the area approximation would improve and approach the Benford probability $\log(1 + \frac{1}{d})$. The stripes are of constant width; so, the only way for there to be more stripes is for the curve to be wider, spanning more integers. Each integer represents an order of magnitude of the data x.

In summary, data distributions will conform to Benford's law if they span many orders of magnitude and if the distribution curve is reasonably smooth. Constrained distributions such as human body temperatures and heights of adult males are smooth, but not Benford-like because they span no more than one order of magnitude. Better candidates include numbers randomly chosen from a newspaper or from a list of physical constants, where the data is almost certain to range over several orders of magnitude. Hill's distribution of distributions, formed from multiple data sources, would obviously be wide enough to satisfy Benford's law.

What Good Is a Newborn Baby?

In a letter dated July 21, 1978, to mathematician Ralph Raimi, Dutch-American physicist Samuel Goudsmit wrote [Goudsmit 78, p. 218],

> To a physicist Simon Newcomb's explanation of the first-digit phenomenon is more than sufficient "for all practical purposes." Of course here the expression "for all practical purposes" has no meaning. There are no practical purposes, unless you consider betting for money on first-digit frequencies with gullible colleagues a practical use. In physics, and in much of mathematics, proofs and derivations have a direct bearing on further basic developments and new problems. The first digit phenomenon has no such usefulness, it stands alone as an intriguing diversion.

What good is a newborn baby?

Almost a century passed before Benford's law was put to work. Applications include the testing of mathematical models, computer design, and the detection of accounting fraud.

Benford-in, Benford-out

The intuitively unexpected Benford distribution of first digits is, in fact, expected behavior for a variety of data types. This being so, mathematical models intended to predict future stock indices, census data, and other types of data having wide ranges can be tested. If current data conforms to Benford's law, then predicted data should do the same. If the data does not conform, the model itself may be defective.

Computer Design

If data used for scientific calculations is expected to conform to Benford's law, then computers can be designed to efficiently accommodate this data. (File sizes on personal computer hard drives also conform.) Consideration is given to the Benford proportions so as to minimize storage (disk) space and maximize rate of output. Knowledge of stored data may also improve virus and error detection.

Hill gives a cash register analogy [Hill 98]. If the frequency of denominations of bills used for transactions is known beforehand, the register drawer can be designed so as to accommodate the frequently used denominations by making the bins of different depths and placing them in a convenient arrangement, such as the keys of a computer keyboard.

Fraud Detection

Establishing guilt in the US criminal justice system is somewhat analogous to statistical hypothesis testing. *Ei incumbit probatio, qui dicit, non qui negat; cum per rerum naturam factum negantis probatio nulla sit.* The proof lies upon him who affirms, not upon him who denies; since, by the nature of things, he who denies a fact cannot produce any proof. In statistical hypothesis testing a claim is made, evidence (data) is collected, and a conclusion is drawn about the claim based on the evidence. There are two hypotheses—the null hypothesis (H_0) and the alternative hypothesis (H_1). The null hypothesis represents the default position and is assumed true until or unless significant evidence suggests otherwise; if so, it is rejected and the alternative hypothesis—a different position—is supported. Think of the null hypothesis as the *status quo* hypothesis and the alternative hypothesis as the challenge. One never *proves* either hypothesis. Ultimately, the conclusion will be to either reject the null hypothesis or fail to reject it at a specified statistical level of significance.

In a criminal trial, the defendant is assumed innocent until significant evidence is presented suggesting the defendant is guilty. (For criminal trials, the word *significant* is taken to mean *beyond reasonable doubt.* In a civil trial, only a *preponderance of evidence* is required for the plaintiff to win the case.) The assumption of innocence is analogous to the null hypothesis and guilt becomes the alternative hypothesis. Ultimately the jury either finds the defendant guilty (supports the alternative hypothesis) or acquits the defendant (fails to support the alternative hypothesis). As is the case with statistical hypothesis testing, we can't be 100 percent confident that a verdict is correct. A jury may fail to convict a guilty defendant and defendants are sometimes convicted for crimes that they did not commit. For various forms of fraud, Benford's law serves as a forensic tool that may produce incriminating evidence.

Data fabrication may be detectable simply because it's hard to fake random data. Hill demonstrates this to his students by giving them the following exercise as a homework assignment. Half the class is asked to flip a coin 200 times and record the results (heads or tails). The other half is asked to fake such results by writing down a randomly appearing sequence of heads and tails. When students return to class with their assignments, Hill spots the fake coin toss sequences with 95 percent accuracy [Hill 98]. Recalling the *hot hand fallacy* of Chapter 4, random sequences are far *streakier* than one might expect. For a true sequence of 200 random tosses, it is highly likely for there to be a run of six consecutive heads or six consecutive tails. Someone faking the data would likely not include runs of this length and are easily spotted as fakers.

Falsified data may not show the Benford distribution. If a large random data set is expected to conform to the Benford distribution and fails to do so, one might suspect the data is not truly random. Data failing to conform does not, in and of itself, prove fraud any more than a defendant's blood type or finger prints matching those found at the crime scene proves the defendant is guilty. It is, however, evidence of an irregularity that may warrant further investigation.

Statistical methods, including Benford's law, have been used to detect the falsification of clinical trials data. Such fraud is rare; however, the consequences may be severe, with potential harm to patients, researchers, and pharmaceutical companies. The incentives for falsifying data include financial gain, professional advancement, and just plain laziness.

Fraudulent manipulation may involve *trimming* (reducing the variance and preserving the mean by deleting outliers), *cooking* (picking and choosing which data to keep and which to delete), and flat out *forgery* (falsification). If the data is fraudulent, statistical methods may detect the following irregularities:

1. abnormal digit frequencies (detectable by Benford's law),
2. too small or too large variances,
3. irregular distribution/skewedness,
4. too few or too many outliers,
5. irregularities associated with rounding.

As noted, statistical tests alone can not prove fraud. In fact, fraud may be the least likely explanation of a detected irregularity. Extensive exploration may yield false positives and cleverly manipulated data may slip through as a false negative. And, fraudulent data may not significantly alter the study's conclusion. Nevertheless, Benford's law and other statistical tests may be of great value in minimizing fraud in clinical trials.

Election fraud (vote count falsification) may have wider implications affecting an entire electorate. Statistical tests, including Benford's law, rely on the tester's access to vote counts and other election data. Such tests have been used to test the legitimacy of the 2004 Venezuelan presidential referendum, the 2006 Mexican presidential election, and the 2009 Iranian presidential election.

In August of 2004 Venezuelans voted in a presidential recall referendum, deciding if President Hugo Chávez should remain in office. Chávez survived the challenge with 60 percent of the electorate voting that he fulfill his term. To legitimize the outcome, two audits were conducted and supervised by the Organization of American States (OAS) and the Carter Center, a nongovernmental human rights organization founded by former US President Jimmy Carter. Nevertheless, the outcome was challenged by the opposition, with claims of fraud relating to electronic voting. Failure of the vote totals to conform with Benford's law in the first and second digits was given as evidence.

Why consider second digits? At precinct-level, legitimate first digit data may not be expected to conform to Benford's law. To see why, assume precinct sizes are similar, each containing approximately 1,000 voters. In a close race between two candidates, each would receive approximately 50 percent of the vote and precinct vote counts would almost all have leading digits of 4 or 5, as opposed to a Benford-like distribution. The second digit may be a better choice when using Benford's law to test for vote count fraud.

An independent panel formed by the Carter Center rejects the applicability of Benford's law [Carter Center 05, pp. 132–134]:

> The panel believes there are many reasons to doubt the applicability of Benford's Law to election returns. In particular, Benford's Law is characteristic for scale-invariant data, while election machines are allocated to maintain a relatively constant number of votes per machine. ... In short, Benford's Law does not generally apply to electoral data and even in cases where we suspect that it might apply, we find that it does not. All in all, Benford's Law seems like a very weak instrument for detecting voting fraud. ... Alleged evidence of fraud based on Benford's Law instead demonstrates how closely the election data match reasonable models of the election.

Overall, the Center notes some irregularities in the Venezuelan recall referendum but fails to find significant evidence of fraud and believes the election result reflects the will of the electorate.

Ricardo Mansilla of the Universidad Nacional Autónoma de México (UNAM) applies a Benford analysis to the two main candidates of the 2006 Mexican presidential election. The vote counts for Felipe Calderón and Andrés M. López Obrador were analyzed and fail to conform to the Benford distribution. Mansilla conservatively concludes [Mansilla 2006, p. L24],

> It is very difficult to explain why the real data of the voting are how they are, if you compare them with the theoretical results ... and every divergence from Benford's law must be looked at with suspicion.

Political science and statistics professor Walter Mebane of the University of Michigan analyzed data from the 2009 Iranian presidential election, in which Mahmoud Ahmadinejad was declared to have won by a large margin. Using Benford's law and data from Iran's 2005 presidential election, Mebane suspects fraud. His work and updates are posted on his personal website.

Mark Nigrini, Associate Professor of Business at The College of New Jersey, pioneered widespread application of Benford's law to the detection of financial fraud. Inexpensive programs operating with Excel detect and evaluate digital frequency irregularities, flagging possible fraud in tax evasion, campaign finance, insurance claims, and other areas of potential financial fraud. Common *Digital analysis* (DA) tests cover first position digits, second position digits (useful when the first position is not expected to conform to Benford's law), first two positions' digits, first three positions' digits (used to detect abnormally duplicated data), and last two positions' digits (for detecting errors associated with rounding).

DA has become an inexpensive and easy-to-use tool for auditors, both internal and external, to detect accounts and transactions in need of further investigation. Specific applications include interest paid and earned data, accounts payable and receivable data, estimations in the general ledger, and customer refunds.

Nigrini's reputation as a tax fraud sleuth was built in large part on his work with the district attorney's office in Brooklyn, New York. Chief financial investigator Robert Burton, with Nigrini's assistance, successfully detected fraud at seven companies, identified the fraudsters, and charged them with theft [Berton 95, p. B1].

Nigrini works with federal and state tax authorities, and several foreign countries, assisting with tax fraud investigations. He also investigated 13 years of former US President Bill Clinton's tax returns, detecting no red flags with respect to Benford's law [Berton 95, p. B1].

A well-publicized case of financial fraud is that of Kevin Lawrence of Brainbridge Island, Washington. In 2003 Lawrence was sentenced to 20 years in prison and ordered to pay $91 million in restitution for what may be the largest case of stock fraud in Washington State's history. Five thousand investors from 36 states and several foreign countries invested in a chain of health clubs created by Lawrence. Rather than use the money for the business, Lawrence and friends spent most of the money on personal items, attempting a cover up by moving the money through various bank accounts and shell companies. Washington's Department of Financial Institutions, with the assistance of Darrell Dorrell, a forensic accountant, compared 70,000 numbers drawn from various accounts and wire transfers to the Benford distribution. The numbers failed the test [Mlodinow 08, p. 84]. With this and other evidence, Lawrence and his cohorts were convicted.

Scott Petucci , a staff auditor at Utica National Insurance Company, describes how Benford's law can be used to detect irregularities associated with automobile insurance first-party claim fraud. Through automation, fraud detection can be made more efficient, targeting a smaller sample of files for further review. Pettuci concludes [Pettuci 05, p. 34],

> Additional areas of insurance claim payments, other than first-party automobile physical damage, should be researched to determine the effectiveness of using Benford's law to identify number irregularities and ultimately fraudulent claim activity. ... the law may be as applicable, if not more so, to other areas of insurance claim payments.

Caution is required when using DA methods to detect accounting fraud. High costs are associated with false positives (evidence of fraud when none exists) and false negatives (failure to detect existing fraud). As is the case with DNA, fingerprints, and all forms of forensic evidence, no tests are foolproof. In 2009, Bernard Madoff was sentenced in US federal court to 150 years in prison for defrauding investors of billions of dollars using a Ponzi scheme fraud. Data from the Madoff case passes a Benford test and regulators believe sophisticated methods of data falsification were used so as to make the data conform to the Benford distribution.

Appendix

Exponential Data

The exponential growth equation $y(t) = y_0 e^{kt}$, for $k > 0$, models continuous exponential growth with t denoting time and k being the relative rate of increase of the quantity y. For simplicity, assume $y_0 = 1$ and allow t to continuously increase, $0 \leq t < \infty$. If $1 \leq y < 10$, then $0 \leq t < \frac{\ln 10}{k}$. Over this interval, the first digit of y is 1 for $1 \leq y < 2$, corresponding to $0 \leq t < \frac{\ln 2}{k}$. Thus, over this interval, the proportion of t values corresponding to y having a first digit of 1 is given by

$$\frac{\frac{\ln 2}{k}}{\frac{\ln 10}{k}} = \frac{\ln 2}{\ln 10} = \log 2.$$

If $10 \leq y < 100$ then $\frac{\ln 10}{k} \leq t < \frac{\ln 100}{k}$. Over this interval, the first digit of y is 1 for $10 \leq y < 20$, corresponding to $\frac{\ln 10}{k} \leq t < \frac{\ln 20}{k}$. Over this interval, as before, the proportion of t values corresponding to y having a first digit of 1 is given by

$$\frac{\frac{\ln 20}{k} - \frac{\ln 10}{k}}{\frac{\ln 100}{k} - \frac{\ln 10}{k}} = \frac{\ln 2}{\ln 10} = \log 2,$$

the same proportion as for $1 \leq y < 10$. This process can be continued indefinitely, each giving the proportion log 2, as prescribed by Benford's law. Similar calculations for first digits 2 though 9 also agree, precisely, with the Benford expression $\log(1 + \frac{1}{d})$.

Scale Invariance

We can express numerical data in the scientific notation form $a \times 10^n$, where $1 \leq a < 10$ and n is an integer. Further assume the lead digit(s) as given by a are scale invariant. As an example, say the proportion of data corresponding to $1 \leq a \leq 2$ is p. Then, if all data are scaled (multiplied) by some positive quantity c, the proportion of scaled data with first digits in this interval remains equal to p. This occurs whenever $0 \leq \log a \leq \log 2$ and is associated with the same proportion p. Consequently, $\log a$ is scale invariant for $0 \leq \log a < 1$. Since $\log ca = \log c + \log a$, scaling the data effectively translates the graph of the probability density function for $\log a$ to the right or left, depending on the sign of $\log c$. Scale invariance requires that the shape of the graph on $[0,1)$ should not change. (If part of the graph is translated beyond 1, or below 0, the extended part should wrap around and be placed at the opposite end. This behavior is *clock-like*, similar to what occurs when the hours extend past 12 noon: 13, 14, 15, become 1, 2, 3, ...) The only probability density function that will remain unchanged over $[0,1)$ under all such translations is the uniform (flat) distribution, graphed as a horizontal straight line. If this were not the case, then the peaks and valleys of the curve would shift and the shape would change.

Now let's compute the probability of the first digit being 1 for data having scale-invariant first digits. The first digit is 1 whenever $1 \leq a < 2$, corresponding to $0 \leq \log a < \log 2$. To compute the associated probability, we integrate the uniform probability density function between 0 and $\log 2$, yielding

$$\int_0^{\log 2} 1\, dy = \left[y \right]_0^{\log 2} = \log 2,$$

which agrees with Benford's law. In general, the probability that the first digit is d, $d = 1, 2, 3, ...9$, is

$$\int_{\log d}^{\log(d+1)} 1\, dy = \left[y \right]_{\log d}^{\log(d+1)} = \log(d+1) - \log d = \log\left(\frac{d+1}{d} \right) = \log\left(1 + \frac{1}{d} \right),$$

the Benford expression.

Let the Mystery Be!

There ain't no answer.
There ain't going to be any answer.
There never has been an answer.
That's the answer.
—Gertrude Stein

T here are no answers to be found at the back of this book, not even "Answers to Selected Exercises":

- Is Pascal's wager reason enough for a rational individual to believe in God?
- Is the St. Petersburg envelope problem resolvable?
- Does Blackwell's technique of choosing the better of two options have real-world applications?
- Will additional significant real-world applications of Parrondo's paradox be discovered?
- How does one choose rationally in an environment of imperfect recall?
- When faced with the choices of Newcomb's paradox, which would be your preference?
- What strategy would you adopt as a player of prisoner's dilemma? Specifically, if you were a contestant on the British TV game show *Golden Balls*, would you "split" or "steal"?
- Will cooperation evolve in an iterated version of prisoner's dilemma?

These are not contrivances. We encounter prisoner's dilemma, imperfect recall, and other paradoxical challenges involving mathematical expectation on a daily basis. So, is it proper to conclude this book with so many unanswered questions?

Proper or not, the book ends here. Twentieth-century German mathematician David Hilbert, the founder of mathematical formalism, issued a general challenge to the mathematical sciences [Hilbert 64, pp. 183–201]. "The goal ... is to establish once and for all the certitude of mathematical methods .... The present state of affairs where we run up against paradoxes is intolerable. ... If mathematical thinking is defective, where are we to find truth and certitude?" His words, *"Wir müssen wissen, wir werden wissen"* are written on his grave: "We must know. We shall know." But, in actuality, there are countless mathematical propositions that are inherently undecidable. And, perhaps the greatest paradox of all is that there are paradoxes in mathematics. So, we may be doomed to failure when it comes to Hilbert's challenge.

Is absence of an answer or lack of resolution a bad thing? Not always. Nobel Prize winning French author André Gide writes [Gide 73, p. 353], "One doesn't discover new lands without consenting to lose sight of the shore for a very long time." So, why not make the best of it and enjoy the adventure?

I've taught mathematics my entire professional life, and it's not my intention to celebrate ignorance; yet, there is something about the unknown that is to be appreciated and for which we must be thankful. Mathematical expectation, though useful, generates its own paradoxes, forcing us to reconsider basic premises. Even if our questions remain unanswered, the surprises are entertaining and the questions are challenging. With or without resolution, we should enjoy the experience.

Knowing some propositions are undecidable and that some mathematical questions may have no universally acceptable answer, what is the point in endlessly pursuing a potentially unattainable goal? My colleague, Carl Broderick, answers honestly, "Because it's fun!"

> Everybody's wonderin' what and where they all came from.
> Everybody's worryin' 'bout where they're gonna go when the whole thing's done.
> But no one knows for certain and so it's all the same to me.
> I think I'll just let the mystery be.
>
> —Iris Dement

I too choose to "let the mystery be."

—Leonard M. Wapner
Division of Mathematical Sciences
El Camino College

Bibliography

Abbott, D. "Developments in Parrondo's Paradox." In In, V., Longhini, P., and Palacios, A. (editors), *Applications of Nonlinear Dynamics: Model and Design of Complex Systems*, Understanding Complex Systems, pp. 307–322. Berlin: Springer-Verlag, 2009.

Aczel, A. *The Mystery of the Aleph: Mathematics, the Kabbalah, and the Search for Infinity.* New York: Four Walls Eight Windows, 2000.

Ameriks, J., Caplin, A., and Leahy, J. "The Absentminded Consumer." Working Paper 10216, National Bureau of Economic Research, Cambridge, MA, 2004.

Aumann, R., Hart, S., and Perry, M. "The Absent-Minded Driver." *Games and Economic Behavior* 20 (1997), 102–116.

—. "The Forgetful Passenger." *Games and Economic Behavior* 20 (1997), 117–120.

Austin, J. "Overbooking Airline Flights." *The Mathematics Teacher* 75:3 (1982), 221–223.

Axelrod, R. *The Evolution of Cooperation.* New York: Basic Books, Inc., 1984.

Axelrod, R. and Dion, D. "The Further Evolution of Cooperation." *Science* 242 (1988), 1389.

Bell, A. E. *Christian Huygens and the Development of Science in the Seventeenth Century.* London: Edward Arnold and Company, 1947.

Benenson, J., Pascoe, J., and Radmore, N. "Children's Altruistic Behavior in the Dictator Game." *Evolution and Human Behavior* 28 (2007), 168–175.

Bennett, C. et al. "Free-Riding and the Prisoner's Dilemma: Problems in Funding Economic Analyses of Phase III Cancer Clinical Trials." *Journal of Clinical Oncology* 13:9 (1995), 2457–2463.

Bernoulli, J. *The Art of Conjecturing*, translated by Sylla, E. D. Baltimore: The Johns Hopkins University Press, 2006. (Originally published in Latin as *Ars Conjectandi* in 1713.)

Berton, L. "He's Got Their Number: Scholar Uses Math to Foil Financial Fraud." *Wall Street Journal*, July 10, 1995, B1.

Blackwell, D. "On the Translation Parameter Problem for Discrete Variables." *The Annals of Mathematical Statistics* 22:3 (1951), 393–399.

Board, O. "The Not-So-Absent-Minded-Driver." Department of Economics Discussion Paper 147, University of Oxford, Oxford, UK, 2003. (Available at http://www.economics.ox.ac.uk/Research/wp/pdf/paper147.pdf.)

Brams, S. J. *Superior Beings: If They exist, How Would We Know?* New York: Springer-Verlag, 1983.

Brams, S. J. and Kilgour, D. M. "The Box Problem: To Switch or Not to Switch." *Mathematics Magazine* 68:1 (1995), 27–34.

Brosnan, S. and de Waal, F. B. M. "Monkeys Reject Unequal Pay." *Nature* 425 (2003), 297–299.

Cardano, G. "Liber de Ludo Aleæ (The Book on Games of Chance)," translated by Gould, S. H. In Ore, Ø., *Cardano, The Gambling Scholar*, pp. 181–241. Princeton: Princeton University Press, 1953. (Originally published in Latin in 1663.)

—. *The Book of My Life (De Vita Propria Liber)*, translated by Stoner, J. New York Review Books: New York, 2002. (Originally published in Latin in 1643.)

The Carter Center. "Observing the Venezuela Presidential Recall Referendum: Comprehensive Report." Available at http://www.cartercenter.org/documents/2020.pdf, 2005.

Cicero, M. Quoted in Weaver, W., *Lady Luck: The Theory of Probability*, p. 53. New York: Dover Publications, Inc., 1963.

Dawkins, R. *The God Delusion*. London: Random House, 2006.

Efron, B. and Morris, C. "Stein's Paradox in Statistics." *Scientific American* 236:5 (1977), 119–127.

Elga, A. "Self-Locating Belief and the Sleeping Beauty Problem." *Analysis* 60:2 (2000), 143–147.

Ellsberg, D. "Risk, Ambiguity, and the Savage Axioms." *The Quarterly Journal of Economics* 75:4 (1961), 643–669.

Everson, P. "Stein's Paradox Revisited." *Chance* 20:3 (2007), 49–56.

Fewster, R. M. "A Simple Explanation of Benford's Law." *The American Statistician* 63:1 (2009), 26–32.

Flehinger, B. J. "On the Probability that a Random Integer has Initial Digit A." *The American Mathematical Monthly* 73:10 (1966), 1056–1061.

Fox, C. and Tversky, A. "Ambiguity Aversion and Comparative Ignorance." *The Quarterly Journal of Economics* 110:3 (1995), 585–603.

Franklin, J. *The Science of Conjecture: Evidence and Probability before Pascal.* Baltimore: The Johns Hopkins University Press, 2001.

Gale, D. *Tracking the Automatic Ant and Other Mathematical Explorations.* New York: Springer-Verlag, 1998.

Gardner, M. "Free Will Revisited with a Mind-Bending Paradox by William Newcomb." *Scientific American* 229:1 (1973), 104–109.

—. *Aha! Gotcha: Paradoxes to Puzzle and Delight.* New York: W. H. Freeman and Company, 1982.

—. *Knotted Doughnuts and Other Mathematical Entertainments.* New York: W. H. Freeman and Company, 1986.

—. *The Night is Large: Collected Essays, 1938–1995.* New York: St. Martin's Press, 1996.

Gardner, M. and Nozick R. "Reflections on Newcomb's Problem: A Prediction and Free Will Dilemma." *Scientific American* 230:3 (1974), 102–108.

Gelman, A., King, G., and Boscardin, J. "Estimating the Probability of Events That Have Never Occurred: When is Your Vote Decisive?" *Journal of the American Statistical Association* 93:441 (1998), 1–9.

Gide, A. *The Counterfeiters.* New York: Vintage Books, 1973.

Gilboa, I. and Gilboa-Schechtman, E. "Mental Accounting and the Absentminded Driver." In Brocas, I. and Carrillo, J. D. (editors), *The Psychology of Economic Decisions, Volume 1: Rationality and Well-Being*, pp. 127–136. New York: Oxford University Press, 2003.

Gilligan, L. and Nenno, R. *Finite Mathematics with Applications to Life*, Second Edition. Santa Monica: Goodyear Publishing Company, Inc., 1979.

Gilovich, T. *How We Know What Isn't So: The Fallibility of Human Reason in Everyday Life.* New York: The Free Press, 1991.

Gneezy, U. and Rustichini, A. "A Fine is a Price." *Journal of Legal Studies* 29:1 (2000), 1–17.

Goudsmit, S. Quoted in Raimi, R., "The First Digit Phenomenon Again," *Proceedings of the American Philosophical Society* 129:2 (1985), 211–219.

Guillen, M. *Five Equations that Changed the World: The Power and Poetry of Mathematics.* New York: Hyperion, 1995.

Hacking, I. *The Emergence of Probability: A Philosophical Study of Early Ideas about Probability, Induction, and Statistical Inference*, Second Edition. New York: Cambridge University Press, 2006.

Hardin, G. "The Tragedy of the Commons." *Science* 162 (1968), 1243–1248.

Harmer, G. and Abbott, D. "Losing Strategies Can Win by Parrondo's Paradox." *Nature* 402 (1999), 864.

Hilbert, D. "On the Infinite." In Benacerraf, P. and Putnam, H. (editors), *Philosophy of Mathematics: Selected Readings*, pp. 183–201. Englewood Cliffs: Prentice Hall, 1964.

Hill, T. P. "A Statistical Derivation of the Significant-Digit Law." *Statistical Science* 10:4 (1995), 354–363.

—. "The First Digit Phenomenon." *American Scientist* 86:4 (1998), 358–363.

Hoyle, F. *The Intelligent Universe: A New View of Creation and Evolution.* London: Michael Joseph Limited, 1983.

Huygens, C. *Christiani Hugenii Libellus de Ratiociniis in Ludo Aleae; or The Value of All Chances in Games of Fortune; Cards, Dice, Wagers, Lotteries, etc. Mathematically Demonstrated,* translated by Browne, W. London: printed by Keimer, S., for Woodward, T., 1714. Electronic reproduction available, Thomson Gale, Farmington Hills, MI, 2003. (Originally published in Latin as *De Ratiociniis in Ludo Aleae* in 1657.)

Ijiri, Y. and Leitch, R. "Stein's Paradox and Audit Sampling." *Journal of Accounting Research* 18:1 (1980), 91–108.

Kahneman, D. and Tversky, A. "Prospect Theory: An Analysis of Decision under Risk." *Econometrica* 47:2 (1979), 263–292.

Kilgour, D. M. and Brams, S. J. "The Truel." *Mathematics Magazine* 70:5 (1997), 315–325.

Kraitchik, M. *Mathematical Recreations,* Second Edition. New York: Dover Publications, Inc., 1953.

Lakshminaryanan, V., Chen, M. K., and Santos, L. R. "Endowment Effect in Capuchin Monkeys." *Philosophical Transactions of The Royal Society* 363 (2008), 3837–3844.

Lec, S. Quoted in Galazka, J., *A Treasury of Polish Aphorisms,* p. 35. New York: Hippocrene Books, Inc., 2002.

Lewis, D. "Prisoners' Dilemma Is a Newcomb Problem." *Philosophy and Public Affairs* 8:3 (1979), 235–240.

Lipman, B. "More Absentmindedness." *Games and Economic Behavior* 20 (1997), 97–101.

Lorenz, E. *The Essence of Chaos.* Seattle: University of Washington Press, 1993.

Luce, R. D. and Raiffa, H. *Games and Decisions.* New York: Dover Publications, Inc., 1985.

Luenberger, D. *Investment Science.* New York: Oxford University Press, 1997.

Mansilla, R. Quoted in Torres, J., Fernández, S., Gomero, A. and Sola, A., "How Do Numbers Begin? (The First Digit Law)," *European Journal of Physics* 28:3 (2007), L17–L25.

McKean, K. "Decisions, Decisions." *Discover* 6:6 (1985), 22–31.

Mlodinow, L. *The Drunkard's Walk: How Randomness Rules Our Lives.* New York: Pantheon Books, 2008.

Nash, J. "The Bargaining Problem." *Econometrica* 18 (1950), 155–162.

Nozick, R. "Newcomb's Problem and Two Principles of Choice." In Rescher, N. (editor), *Essays in Honor of Carl G. Hempel,* pp. 114–146. Dordrecht: D. Reidel Publishing Company, 1969.

Pascal, B. *Pensées (Thoughts)*, translated by Trotter, W. F. Stilwell, KS: Digireads.com Publishing, 2005.

Paulos, J. *A Mathematician Plays the Stock Market.* New York: Basic Books, 2003.

Petucci, S. D. "Benford's Law: Can It Be Used to Detect Irregularities in First Party Automobile Insurance Claims?" *Journal of Economic Crime Management* 3:1 (2005), 1–35.

Piccione, M. and Rubinstein, A. "On the Interpretation of Decision Problems with Imperfect Recall." *Games and Economic Behavior* 20 (1997), 3–24.

Pinkham, R. "On the Distribution of First Significant Digits." *The Annals of Mathematical Statistics* 32:4 (1961), 1223–1230.

Pisano, G. P. "Can Science Be a Business? Lessons from Biotech." *Harvard Business Review* 84:10 (2006), 114—124.

Polgreen, P. "Nash's Arbitration Scheme Applied to a Labor Dispute." *UMAP Journal* 13 (1992), 25–35.

Rapoport, A. *Fights, Games, and Debates.* Ann Arbor: The University of Michigan Press, 1974.

Resnik, M. *Choices: An Introduction to Decision Theory.* Minneapolis: University of Minnesota Press, 1987.

Russell, B. W. *Common Sense and Nuclear Warfare.* London: George Allen and Unwin, 1959.

vos Savant, M. *The Power of Logical Thinking: Easy Lessons in the Art of Reasoning and the Hard Facts about Its Absence in Our Lives.* New York: St. Martin's Griffin, 1996.

Scarne, J. *Scarne's New Complete Guide to Gambling.* New York: Simon and Schuster, 1974.

Schaeffer, J. et al. "Checkers Is Solved." *Science* 317 (2007), 1518–1522.

Shannon, C. "Programming a Computer for Playing Chess." *Philosophical Magazine* 41 (1950), 256–275.

Snell, J. L. (editor). *Topics in Contemporary Probability and Its Applications.* Boca-Raton: CRC Press, 1995.

Stjernberg, F. "Parrondo's Paradox and Epistemology—When Bad Things Happen to Good Cognizers (and Conversely)," *Hommage à Wlodek: Philosophical Papers Dedicated to Wlodek Rabinowicz.* Available at http://www.fil.lu.se/hommageawlodek/site/papper/ StjernbergFredrik.pdf, 2007.

Straffin, P. *Game Theory and Strategy.* Washington, DC: The Mathematical Association of America, 1993.

Tom, S., Fox, C., Trepel, C., and Poldrack, R. "The Neural Basis of Loss Aversion in Decision-Making under Risk." *Science* 315 (2007), 515–518.

Tversky, A. and Koehler, D. "Support Theory: A Nonextensional Representation of Subjective Probability." *Psychological Review* 101:4 (1994), 547–567.

Wapner, L. *The Pea and the Sun: A Mathematical Paradox.* Wellesley, MA: A K Peters, Ltd., 2005.

Waugh, K. et al. "A Practical Use of Imperfect Recall." In Bulitko, V. and Beck, J. C. (editors), *Proceedings of the Eighth Symposium on Abstraction, Reformulation, and Approximation*, pp. 175–182. Palo Alto, CA: AAAI Press, 2009.

Yandell, B. *The Honors Class: Hilbert's Problems and Their Solvers.* Natick, MA: A K Peters, Ltd., 2002.

Index

A

Abbott, Derek, 112
absentminded driver, 128–130
addition rule
   general, 25
   mutually exclusive events, 25
airline overbooking, 47–50
ambiguity aversion, 71, 75–77
arbitration polygon, 163–167
Aristotle, 1
Ars Conjectandi (Art of Conjecture), 11
Ask Marilyn, 100
astragalus bone, 2
Aumann, Robert, 131
availability error, 82
average of averages, 182
aversion, 71–89
   ambiguity, 75–77
   inequality, 78
   loss, 72–74
   risk, 78–79, 82–89
Axelrod, Robert, 158
axiomatic system, 14

B

Bashevis, Isaac, 169
Bay of Pigs, 149
being in the zone, 86
Benford, Frank, 177
Benford's law, 172, 177–190
   data conforming, 180
   Newcomb's discovery, 178
Bernoulli, Daniel, 64
   approach to St. Petersburg problem, 95
   Law of Hydrodynamic Pressure, xiv
Bernoulli, Jacob, 10–12
Bernoulli, Nicholas, 63
Bernoulli family, 11
Bernoulli's law, 10–13, 21
   Hydrodynamic Pressure, xiv
   large numbers, 42, 85
Bernoulli trials. *See* binomial
   probabilities
Berra, Yogi, 99
binomial probabilities, 31
   formula, 32
binomial trials, 31

Blackwell, David, 97
Blackwell's bet, 97–98
Bohr, Niels, 99
bones (dice), 2
Brams, Steven, 94
Brazil nut effect, 116
Broderick, Carl, 192
Brownian motor, 110
Buffet, Jimmy, 104
bumping, 47
butterfly effect, 2
bystander effect, 152

C

calculus of variations, 120
Calvinism, 174
Cardano, Girolamo, 3–5
cardinality, 19
Carter, Jimmy, 186
Carter Center, 187
Chávez, Hugo, 186
checkers solvability, 68–69
chess solvability, 68–69
chicken, 146–153
Clinton, Bill, 188
Cold War, 149
Cold Within, The (Kinney), 153
complement, 23–24
composite sampling, 51–53, 68
compound event, 19
conditional probabilities, 26
confusion of inverse, 27
convex hull, 143
cooking (data), 186
Cornwall mines, 110–111
Cox, Ronny, 154
Cuban missile crisis, 149
*Curious Incident of the Dog in the
    Night-Time, The* (Haddon), 102

D

Dawkins, Richard, 55, 124
Dean, James, 147
deceit aversion, 71
Deliverance, 154
Dement, Iris, 192

de Méré, Chevalier, 7
De Moivre, Abraham, 13
De Ratiociniis in Ludo Aleae (Calculating
    in Games of Chance) (Huygens), xiii, 8,
    13
Descartes, René, 8
dice, 23–24
    all possible results, 5
dictator game, 78–79
digital analysis (DA) tests, 188
diminishing marginal utility, 64
diminishing returns, 35
distribution of distributions, 182
Doctrine of Chances (De Moivre), 13
dominance principle, 172
Dorrell, Darrell, 188

E

effectively impossible, 31
Efron, Bradley, 67
eikos, 3
Einstein, Albert, xiv
Elga, Adam, 134
Ellsberg, Daniel, 76
empirical probability, 20
envelope problem
    double or half, 92–93
    powers of three, 95–96
    St. Petersburg, 94
equiprobable, 22
Euler, Leonhard, 14
event definition, 19
Everson, Phil, 67
exchange conditions, 105
expected value, xiii, 19
    definition, 32–34
    random variable, 33
experimental probability, 20
experiment, definition, 19
exponential data, 189

F

fair game, 60
Faraday, Michael, xiv
Fermat, Pierre de, 6–7
Fewster, R. M., 183

Fewster's stripes, 183
Five Equations That Changed the World
   (Guillen), xiv
flashing ratchet, 110
Flehinger, B. J., 181, 182
forgery, 186
formulating probability, 5
Franklin, James, 3
fraud detection, 185
free rider dilemma, 159
Friend or Foe, 159

G

Gale, David, 95, 104, 105
Gambaud, Antoine, 7
gambler's fallacy, 85
gambling, 4, 41–44
games. *See also* specific name, e.g., trust
   game
   non-zero-sum, 141–168
   with saddle points, 60
   without saddle points, 61
   zero-sum, 56, 141
game theory, 56–62
   first publication, 8–9
Gamow, George, 124
Gardner, Martin, 172
Gauss, Carl Friedrich, 13
General Exchange Condition, 94
general multiplication rule, 30
Genovese effect, 152
Gide, André, 192
Gilboa, Itzhak, 137
Gilboa-Schechtman, Eva, 137
Gilovich, Thomas, 84
Gödel, Kurt, 15
Golden Balls, 159
Golden Rule, 78
Goodman, Steve, 104
Goudsmit, Samuel, 184
Guillen, Michael, xiv
Güth, Werner, 80

H

Haddon, Mark, 102
Hardin, Garrett, 160

Harmer, Gregory, 112
harmonic series, 39
Hart, Sergiu, 131
Hesse, Hermann, 71
Hilbert, David, 14, 15, 192
Hill, Theodore, 182, 183
Horace, 41
Horologium Oscillatorium, 10
hot-bat hitting streaks, 87
hot-hand or hot-streak fallacy, 86
*How We Know What Isn't So: The
   Fallibility of Human Reason in
   Everyday Life* (Gilovich), 84
Hoyle, Fred, 115
Hoyle's Fallacy, 124–125
Huygens, Christiaan, xiii, xiv, 8–10

I

imperfect information, 127
imperfect recall, 127–139
   absentminded driver, 128–130
   applications, 137–139
   Sleeping Beauty problem, 134–136
   unexpected lottery payoffs, 131–133
independence of irrelevant alternatives
   (IIA) condition, 165
inequity aversion, 71, 78
inferential statistics, 65
infinite series, 37–38
information sets, 134
insurance, 45–46
irrational numbers, 16
Irving, John, 127

J

James-Stein estimators, 65–66, 67

K

Kahneman, Daniel, 71, 72
Kennedy, John F., 149, 150
Khrushchev, Nikita, 149, 150
Kilgour, D. Marc, 94
Kinney, James Patrick, 153
Knight, Frank, 76
Knightian uncertainty, 76

Knotted Doughnuts and Other
Mathematical Entertainment (Gardner),
    172
Kolmogorov, Andrei, 13–17
Kolmogorov axioms, 16–17, 21, 25
Kraitchik, Maurice, 91

L

Laplace, Pierre-Simon De, 14
law of averages, 85
Law of Electromagnetic Induction, xiv
Law of Hydrodynamic Pressure, xiv
law of large numbers, 12, 24, 42
law of probability, 22–30
law of small numbers, 85
Lawrence, Kevin, 188
laws of chance, xiii, 4
left-right game, 58–60
Let's Make a Deal, 91, 99, 104
Liber de Ludo Aleae (Book on Games of
    Chance), 4
Lipman, Barton, 131
Lorenz, Edward, 2
loss aversion, 71, 72–74
lottery payoffs, 131–133
Luenberger, David, 116

M

mamihlapinatapei, 153–154
Mansilla, Ricardo, 187
Markov chain, 123
Martingale betting system, 43
mathematical expectation. *See* expected
    value
mathematical theory, 14
measure theory, 14–16
medical diagnostic testing, 28
mental accounting, 137
Mersenne, Marin, 8
Mine, Levant, 110
Monty Hall problem, 99–103
Morgenstern, Oskar, 56
Morris, Carl, 67
Muesli effect, 116
multiplication rule
    general, 30
    independent events, 30
    mutually exclusive events, 24

N

Nash, John, 164
    arbitration scheme, 163–166
nature vs. nurture, 10
Newcomb, Simon, 177, 178
Newcomb, William, 172, 174, 177
Newcomb's paradox, 169–177, 172
    dominance vs. expectation, 170–173
    expected value principle, 173
    prisoner's dilemma, 174–176
Newton, Isaac, xiv
nickel-quarter coin game, 57
Nixon, Richard, 141
nodes, 132
non-equiprobable probability
    distribution, 93
non-zero-sum games, 141–168
    chicken, 146–153
    Nash arbitration scheme, 163–166
    pizza or pâté, 142–144
    prisoner's dilemma, 154–162
    threat, 145
Nozick, Robert, 172

O

optimal strategy, 56

P

Pacioli, Luca, 6
Packel, Edward, 19
panspermia, 124
paradox of omniscience, 149
Paradox of the Neckties (Kraitchik), 91
pareidolia (apophenia, patternicity), 89
Parrondo profits, 116–117
Parrondo's paradox, 109–126
    main engines of Cornwall mines,
        110–111
    physical models, 115–116
    ratchets, 109
    reliabilism, 114
    truels, 117–119
"Parrondo's Paradox and Epistemology"
    (Stjernberg), 114
Pascal, Blaise, xiii, 6–7

Pascal's wager, 54–55
payoff polygon, 143
Perry, Motty, 131
Pinkham, Roger, 181
Pisano, Gary, 162
pizza or pâté, 142–144
Porter, Cole, 79
power set, 20
powers of three envelope problem, 95–96
predictive value, 28
primordial soup, 115
prisoner's dilemma game, 154–162
    Newcomb's paradox, 174
    simplified, 156
    Tucker's formulation, 156
probabilis, 3
probability
    abiogenesis, 115
    and death, 83
    definition, 20–21
    density function, 183
    dice, 23
    distribution, 93
    experimental or empirical, 20
    first textbook, 8
    law, 22–30
    and measure, 16
    subjective, 82–89
    true theoretical, 20
    problem. *See also* specific name, e.g.,
        Sleeping Beauty problem
    division of stakes, 6
    points, 6
process reliabilism, 114
"Prospect Theory," 72

R

Raimi, Ralph, 184
random variable, 33
random walk, 99
Rapoport, Anatol, 34
ratchets, 109–110
rational numbers, 16
Rebel Without A Cause, 147
regression to mean, 84
reliabilism, 114
reservations and expected revenue, 49

Return to Paradise, 154
ride for free, 162
risk aversion, 72
risk-seeking behavior, 72
rochambeau, 61, 62
rock-paper-scissors, 61, 62
Russell, Bertrand, 147

S

saddle points, 60
sample points, 19
sample space, 4, 19
scale invariance, 181, 189
Scarne, John, 44
Science of Conjecture, The (Franklin), 3
Second Law of Thermodynamics
    (Clausius), xiv
semiplena probatio (half-proof), 3
sensitivity, 27
sequential truel, 118
Shafted, 159
shrinkage factor, 66
simple event, 19
Sleeping Beauty problem, 134–137
Snell, J. Laurie, 98
Snell-Vanderbei technique, 99
Social Security protection pledge, 151
solution, definition, 56
solvable, 56
South Point to North Point, 120–123,
    125–126
specificity, 27
St. Petersburg envelope problem, 94
St. Petersburg paradox, 63
state distribution vector, 124
Stein, Gertrude, 191
Stein's paradox, 64–67
stick or switch dilemma, 99
Stills, Stephen, 177
Stjernberg, Fredrik, 114
strong law of large numbers, 12
subjective probability, 21, 82–89
*Summa de arithemetica, geometria,
    proportioni e proportionalità* (Pacioli),
    6
sunk-cost dilemmas, 74
switch dilemma, 99

T

talus bone, 2
taxpayer protection pledge, 151
Texas Hold'em, 138
Texas Sharpshooter Effect, 88
theoretical probabilities, 20
Théorie Analytique des Probabilitiés
  (Laplace), 14
Theory of Games and Economic Behavior
  (von Neumann), 56
Theory of Special Relativity Equation, xiv
threat and non-zero-sum games, 145
tick-tack-toe solvability, 68–69
*Tracking the Automatic Ant and Other
  Mathematical Explorations*, 95
tragedy of commons, 160
trimming (data), 186
truel, 117–118, 125
true probabilities, 20
trust game, 80–81
Tucker, Albert, 155
Tversky, Amos, 71, 72
two-by-two game with no saddle point,
  62–63
two losing games, 113
two-person game, 56

U

ultimatum game, 80
Universal Law of Gravity, xiv
utility, 35–36

V

Vanderbei, Robert, 98
van der Rohe, Mies, 91
Van Rekiningh in Spelen van Geluck, 8
van Schooten, F., 8
Venn diagram, 16
volatility pumping, 116, 117
von Neumann, John, 56
vos Savant, Marilyn, 100

W

wager consequences, 54
Wallet Game, 91
weak law of large numbers, 12
win-win, 104, 106–107

Z

Zeilberger, Doron, 109
zero-sum game, 56, 141

Printed in the United States
by Baker & Taylor Publisher Services